Distributed Photovoltaic Grid Transformers

Hemchandra Madhusudan Shertukde, PhD

Distributed Photovoltaic Grid Transformers

CRC Press
Taylor & Francis Group
Boca Raton London New York

CRC Press is an imprint of the
Taylor & Francis Group, an **informa** business

CRC Press
Taylor & Francis Group
6000 Broken Sound Parkway NW, Suite 300
Boca Raton, FL 33487-2742

First issued in paperback 2017

Version Date: 20140131

ISBN 13: 978-1-4665-0581-0 (hbk)
ISBN 13: 978-1-138-07384-5 (pbk)

Library of Congress Cataloging-in-Publication Data

Shertukde, Hemchandra Madhusudan.
    Distributed photovoltaic grid transformers / author, Hemchandra Madhusudan Shertukde.
      pages cm
    Includes bibliographical references and index.
    ISBN 978-1-4665-0581-0 (hardback)
    1. Electric inverters. 2. Photovoltaic power systems--Equipment and supplies. I. Title.

TK7872.I65S54 2014
621.31'26--dc23   2013048945

**Visit the Taylor & Francis Web site at
http://www.taylorandfrancis.com**

**and the CRC Press Web site at
http://www.crcpress.com**

Dedication

To my dear wife Rekha; my lovely, accomplished kids Dr. Amola, Karan, and

Rohan; and as always, my loyal dog Sheri, who always guards my work.

Contents

Preface

The world of alternate energy and the diverse fields of applications it provided served as impetus for me to start contemplating writing a book on distributed photovoltaic (DPV) grid transformers and related applications in fall 2008. That year we were working on a grant received from the Department of Energy (DOE) at the University of Hartford, Connecticut, to study the economic value stream generated by solar energy in the state of Connecticut. As one of the principal investigators of the technical aspects related to power systems and electrical power transformers used in harvesting solar energy, I took up the challenge to address the issues related to such transformers.

Subsequently, I applied for my third sabbatical leave in fall 2011 and was granted one to write a book entitled *Distributed Photovoltaic Grid Transformers*. This sabbatical application to write a book received excellent support from the word "go" from its publisher CRC Press, a Taylor and Francis Company. I am grateful to the University of Hartford for granting me the sabbatical for fall 2012 to accomplish this daunting task, especially in an area as virgin as applications of transformers in the solar field. This book provides a bird's-eye view of the salient aspects of DPV grid transformers and the related applications in a concise and organized manner.

In their lives, professors constantly learn from interactions they have with their students. I am no exception. Jorge Kuljis has always motivated me to achieve the best. Appendix A in this book is my partial amendment and improvement of the code for three-limb cores from his earlier work on five-limb cores under my guidance for his Master of Engineering thesis.

Acknowledgments

I would like to thank the Sabbatical Committee at the University of Hartford, Connecticut, for awarding me a sabbatical leave in fall 2012 to write this book, *Distributed Photovoltaic Grid Transformers*. In addition, the support of Dean Louis Manzione of the College of Engineering Technology and Architecture (CETA) in this daunting and challenging task is greatly appreciated. Further, the encouragement of our Electrical and Computer Engineering Department Head Professor Saeid Moslehpour and Provost Sharon Vasquez at the University of Hartford is invaluable and priceless.

I am extremely grateful to Nora Konopka, publisher of Engineering and Environmental Sciences at CRC Press, a Taylor and Francis Company, for engaging me in this task and tolerating the delay caused in completing this work due to the injury to my right hand in February 2012 and the subsequent surgery in July 2012, or else this book would have been published a year ago. Further the great support from Kathryn Everett, project coordinator at CRC Press, a Taylor and Francis Company, is much appreciated.

Thanks are due to my friends and extended family at the IEEE-Transformer Committee for their support and encouragement, especially Aleksandr ("Sasha") Levin and the "Big Gang" of the Indian Institute of Technology, Kharagpur (IIT-KGP, EE75), classmates who constantly encourage me to be the best in power engineering.

The undertaking of the writing of this book would not have been accomplished without strong support from my family. I salute my wife, Rekha, the primary "rock of support" behind this project for her tolerance and constant encouragement to complete this book.

The Author

 **Hemchandra M. Shertukde,** SM'92, IEEE, was born in Mumbai, India, on April 29, 1953. He graduated from IIT Kharagpur with B.Tech (High Hons) with Distinction in 1975. He received his MS and PhD degrees in electrical engineering with specialty in controls and systems engineering from the University of Connecticut–Storrs, in 1985 and 1989, respectively. Since 1995, he has been a full professor in the Department of Electrical and Computer Engineering in the College of Engineering, Technology, and Architecture (CETA) at the University of Hartford (West Hartford, Connecticut). Since fall 2011, he has been a senior lecturer at the School of Engineering and Applied Sciences (SEAS), Yale (New Haven, Connecticut). He is the principal inventor of two commercialized patents (US Patent 6,178,386 and US Patent 7,291,111). He has published several journal articles in IEEE Transactions and has written two solo books on Transformers and Target Tracking.

1

Introduction

1.1 Introduction to Distributed Photovoltaic (DPV) Grid Power Transformers

The increase in oil prices over the past several years has encouraged all scientists, engineers, and economists to look for alternative energy sources. One of these is the abundant sun's energy which can be harnessed into reusable electric energy to supplement and eventually be a major factor in the overall energy generation, transmission, and delivery equation to customers.

Wind energy, solar energy, and ocean wave energy have recently become noticeable players in this exchange. While the penetration is the most important aspect of sustaining these alternative energies, considerable research and development work has been dedicated to the ancillary equipment needed for such energies to be efficiently delivered to the end user.

Distributed photovoltaic (DPV) grid transformers (DPV-GTs) that are solar energy converter transformers are gradually increasing in number in the field due to the recent focus on renewable energy sources. These transformers are primarily used as step-up transformers but can be used as step-down transformers as well. In the case of photovoltaic solar power, electrical power is generated by converting solar radiation into direct current electricity using semiconductors that exhibit the photovoltaic effect. Photovoltaic power generation employs solar panels made up of a number of cells containing photovoltaic material.

The DC energy is then converted to one- or three-phase AC power using an inverter. The inverter is subsequently connected to a DPV-GT. This DPV-GT is further connected to a bus that can feed a suitable load. Figure 1.1 illustrates the process of conversion of energy from solar radiation into usable electrical power.

Currently there exist a variety of available industry standards that address many of these design, operation, and maintenance aspects. Some key design, operation, and salient maintenance aspects (see Figure 1.2) to be considered are listed below:

1. Islanding
2. Voltage flicker

1

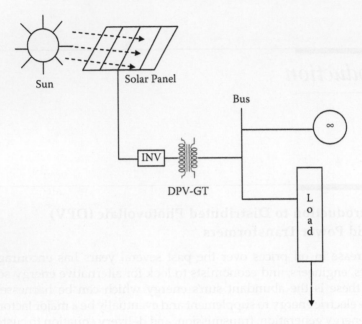

FIGURE 1.1
A solar panel connected to an inverter that is in turn connected to a DPV-GT, a bus, and farther down, a load.

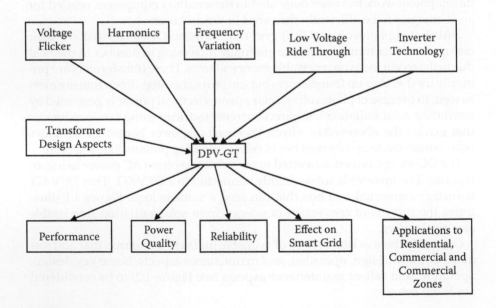

FIGURE 1.2
DPV-GT as a focal point of interest as an application in the alternative solar energy source creation and the effective output characteristics for optimum performance.

3. Voltage operating range

4. Frequency variation

5. Waveform distortion

6. Power factor variation

7. Safety and protection functions

8. Harmonics and waveform distortion

9. Power quality

The objective of this book is to bring to the attention of manufacturers, designers, and users these salient aspects and identify potential gaps as related to other features that comprise the entire understanding of such transformers.

1.2 Voltage Flicker and Variation [10]

Solar transformers operate at a steady voltage, with the rated voltage controlled by inverters. Therefore, voltage and load fluctuations are considerably reduced. The voltage variation is generally in the range of ±5% of the nominal voltage rating. Thus standard design considerations for transformer windings are readily applied from experience. IEEE 519-92 Table 10-2 [10] establishes limits for allowable commutation notch depth introduced by power converters at critical points of the power system, normally coincident with points of transformer allocation.

Load tap changer (LTC) control issues need to be addressed. Some of the LTCs are suitable to the bidirectional power flow, but not all of them.

1.3 Harmonics and Wave Form Distortion [10]

The solar inverter system's typical harmonic content is less than 5%, which has almost no impact on the system (threshold for the normal service conditions). The lower harmonic profile is because there are no generators and switching and protective controls such as those found on wind turbines.

The C57.129 [6] standard sufficiently describes the requirements to the system designer to provide information on the harmonic content and the current waveform, including the cases where there is more than one valve winding on a core leg.

The C57.129 and C57.18.10 standards use a definition of a kVA rating based only on the fundamental frequency. Additional losses due to the harmonic content are taken into account during the heat-run test.

IEEE 519-92 establishes limits for allowable harmonics levels in power systems. Table 10-3 in the standard sets current distortion limits for general distribution systems (120 V through 69,000 V) as a function of the short-circuit ratio and the harmonics order.

In the case of significant harmonic content in DPV-GT, please refer to C57.110, which is an IEEE recommended practice to establish transformer capability when supplying nonsinusoidal load current. Because of today's inverter practice, the harmonics generated from DC to AC conversion may be minimal, but they need to be specified in the customer specifications so that the transformer designer will take account of the additional losses due to the harmonics in the transformer cooling design.

1.4 Frequency Variation [8]

Because the frequency variation can come from the network only, no difference to a "standard" power transformer is expected to be made in its design and manufacture.

1.5 Power Factor (PF) Variation [8]

No significant difference with "standard" power factor practices is expected.

Note standard C57.110 (§5.3 Power factor correction equipment). Power factor correction equipment is frequently installed to decrease utility costs. Care should be taken when this is done, because current amplification at certain frequencies due to resonance in the circuit can be quite high. In addition, the inductance that is reduced in the circuit generally allows higher harmonic currents to exist in the system. Harmonic heating effects from these conditions may be damaging to transformers and other equipment. The additional losses produced may also increase utility costs due to increased wattage requirements, even though the load power factor was improved.

1.6 Safety and Protection Related to the Public [8]

If residential and industrial (nondistributed) PV systems are covered, the safety requirements can have specific features compared to the ones for power transformers, especially when it comes to residential application.

In C57.129, converter transformer pollution aspects are extremely important and should be accurately defined so that proper external insulation (particularly bushings) may be provided.

1.7 Islanding [8]

In these conditions, when the system is functioning but is not connected to the "high-inertia short-circuit capacity" network, then the system could be less stable and subject to frequency variation, but no significant differences with the "standard" transformers are expected. Islanding refers to the condition of a distributed generation (DG) generator continuing to power a location even though power from the electric utility is no longer present. Consider for example a distributed solar facility that has solar panels that feed power back to the electrical grid; in case of a power blackout, if the solar panels continue to power the distributed solar facility, the solar facility becomes an "island" with power surrounded by a "sea" of unpowered distributed solar facilities in a distributed solar grid system. This condition may prove dangerous and sometimes fatal to humans if not properly monitored and controlled.

1.8 Relay Protection [8]

The study of relay protection for DPV-GTs is extremely important given the fact that such transformers operate with inverter circuits between the actual alternative energy source and the eventual connection to such transformers. Such a protective scheme should take into consideration the rapid changes in the conditions both on the source side connected to the PV generator plus the inverter and to the grid side. In PV the amount of sun's rays impinging on the PV panels will be hindered by the presence of clouds and the insolation number for that time of the season.

1.9 DC Bias [10]

In C57.110 (§4.1.4 DC components of load current), harmonic load currents are frequently accompanied by a DC component in the load current. A DC component of load current will increase the transformer core loss slightly but will increase the magnetizing current and audible sound level more substantially. Relatively small DC components (up to the root mean square [RMS] magnitude of the transformer excitation current at rated voltage) are expected to have no effect on the load-carrying capability of a transformer determined by this recommended practice. Higher DC load current components may adversely affect transformer capability and should be avoided.

One of the important parameters that determines how much DC current will cause the saturation of the core is the core construction (three-phase three-limb or three-phase five-limb or single-phase cores). Transformer manufacturers should get the data of possible DC bias current before finalizing the design. Saturation of the core is an important parameter to watch because of the possibility of ferro-resonance in the case of cable-connected pad-mount transformers, because of the possibility of resonance between nonlinear self-inductance of the transformer and other capacitances connected in the system such as cable capacitance and filter capacitance of the inverter under saturation due to DC bias.

1.10 Thermo Cycling (Loading) [8]

In most geographical locations in the United States, solar power facilities experience a steady-state loading when inverters are operating. When the sun comes out, there is a dampened reaction process, and loading on the transformer is more constant. The entire process is controlled by the insolation number in a particular location. The no-load operation of such transformers is completely controlled by a different set of parameters.

Solar power systems typically operate very close to their rated loads. Because load variation from the rated value is appreciably low, the operation of transformers is not adversely affected and does not cause deterioration of parameters that guide the insulation coordination of the core-coil structure. Forces experienced by the primary and secondary windings are not out of the ordinary, thus alleviating problems that may occur in the design of the mechanical structure.

PV system transformers are subject to long-term no-load operation conditions, at least at night. This might have an impact on loss capitalization, which customers usually take into account, and on the transformer design. Storage battery interaction with the transformer in a PV system may control the load consistency and alleviate perceived problems.

The C57.129 standard requires a detailed thermal study, if the transformer or some of the terminals operate above rated capacity. Standard power transformer loading tables should not be used for loading determination because of the effect of the harmonic currents and DC bias on the valve windings (for the high voltage direct current (HVDC) converter transformers).

Even with the loss correction addressing the harmonic content during the heat run, the hot-spot temperature may not be representative of the real conditions due to the nature of the harmonic current distribution in the winding and how it is different during the heat run. "Extended load run with overload" is recommended by CIGRE Joint Task Force 12/14.10-01 (*Electra*, No. 155, August 1994, pp. 6–30) for the HVDC converter transformers.

1.11 Power Quality [8]

Power quality aspects are generally addressed in other sections of this book, although emphasis is on the salient items listed above. The inclusion of several harmonics to the final voltage supplied to the grid can reduce the quality of the sinusoidal signal shape of the power delivered. Thus, filtering is an important aspect to be included in the study of power quality. Commutation of power electronic devices while engaging in improving power quality needs to be studied extensively.

1.12 Low-Voltage Fault Ride-Through

Fault ride-through has yet to be defined for solar systems. This could be because it is easier to turn solar power systems on and off quickly. However, with the advent of smart grid technology, low-voltage fault ride-through has come out of its infancy and a considerable amount of work has been done to alleviate such a condition. This may affect the life of equipment connected to such a system, especially transformers that enable the delivery of AC power to the grid.

1.13 Power Storage [8]

Battery storage impact will depend on the kind of system the DPV-GT is serving in a particular geographic environment. Several ways have been devised to store power by using super capacitors, large batteries, and so forth, but these have limited applications and the end goal of the power required to be generated and delivered will decide the power storage system to be developed.

1.14 Voltage Transients and Insulation Coordination [6]

With solar transformers, step-up duty is required, but without the problems associated with over-voltages caused by unloaded generators. The inverter converts DC input from the photovoltaic array and provides AC voltage to the transformer, giving a steady and smooth transition, with no over-voltage caused by unloaded circuits. All general installations covered under this application have in their system considerations much attention paid to the over-voltage conditions. This problem is addressed by providing an automatic gain control scheme to the inverter circuit configuration.

The C57.129 standard provides specific recommendations on insulation test levels for the converter transformers. The development of a similar recommendation would be appropriate for insulation test levels and procedures required to warrant the reliability of the transformers in the PV application. Figure 1.3 shows a typical core-coil insulation disposition for a related distribution transformer for such DPV grid applications.

1.15 Magnetic Inrush Current [8]

Transformers experience a high current inrush when energized from a de-energized state. The inrush current is typically several times the rated current. The magnitude of the inrush current is determined by a variety of factors defined by the transformer design. The inrush current, when compared as a multiple of rated current, is generally much higher when the energization takes place from the LV side. That is due to the fact that the LV winding is generally the winding that is closest to the core and therefore has a lower air core reactance.

Since the inrush current is several times rated current, each inrush event creates mechanical stresses within the transformer. Frequent energization from a de-energized state should be avoided because it wears down a transformer faster than normal. That can be a consideration for a DPV step-up transformer because the operators could consider saving energy by shutting down the transformers during the night. Such a practice can shorten a transformer's life expectancy.

1.16 Eddy Current and Stray Losses [7]

Eddy currents and stray losses are present in every transformer. The primary stray and eddy losses are due to the 60 Hz fundamental frequency

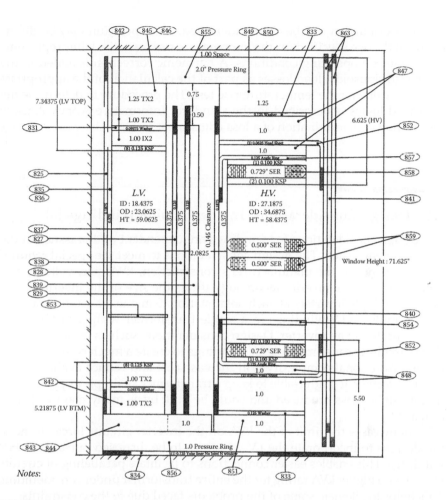

FIGURE 1.3
Typical core-coil insulation disposition for a distribution transformer used in solar applications.

currents. These loss components increase with the square of the frequency and square of the magnitude of the eddy currents. If the inverter feeding the power into the step-up transformer is producing more than the standard level of harmonics, then the stray and eddy losses will increase. The increase in load loss effect on efficiency is not typically a concern. Of much greater concern is the increased hot-spot temperature in the windings and hot spots in metallic parts that can reduce transformer life. A special design transformer can compensate for the higher stray and eddy losses. Also, a larger than necessary kVA transformer can be selected to compensate for higher operating temperatures. However, these concerns on eddy current loss are generally mitigated because the harmonic is less than 1%.

In standard C57.129, the user will clearly indicate the method to be used for evaluating guaranteed load losses. The harmonic spectrum to be used for load loss evaluation will be clearly identified. This spectrum may be different from the one specified for the temperature rise tests; the latter represents the worst-case operating conditions. A harmonic correction is added to the measured sinusoidal load losses as part of the calculation of the appropriate total loss value for the temperature rise test. The procedure to determine the total load losses is described in the standard. Informative Annex A gives the method of determination of a loss adjustment factor.

1.17 Design Considerations: Inside/Outside Windings [6]

The design considerations for windings are dependent on the issues in the previous items listed. The design considerations to meet the special requirements depend on the manufacturer's construction, kVA size, voltage, and other factors. Since inverter technology limits the size of the inverter, there may be multiple inverters at each solar station. Some users would consider having multiple LV windings in a single transformer with each LV winding connected to an inverter. Design considerations such as impedance and short-circuiting cause multiple LV windings to create a much complex transformer. Additional complexity will increase cost and reduce availability of the transformer. It is advisable to keep a transformer as simple as practical so that it can be mass produced and could be built by as many manufacturers as feasible.

In many cases the limit of the kVA on the inverter circuits forces some of the designs to incorporate the LV windings to be disposed outside the HV windings. This enables easier connections to facilitate paralleling of circuits to achieve a higher kVA rating for the entire transformer under consideration. This helps to alleviate some of the problems faced due to the constraints.

1.18 Special Tests Consideration [6]

The following standards are considered: C57.129, "extended load run with overload"; "other power testing concepts and methodologies"; and "specifics for the transformers used with voltage source converters." In addition to the tests, design review is recommended. These transformers fall under a special type where the test conditions are much more stringent. For example IEEE suggests a partial discharge level much higher than some of the requirements as specified by individual end users. These necessitate a higher degree of precaution while designing such transformers, as standard clearances may not serve the purpose as required by the newer end user.

1.19 Special Design Consideration

Solar power systems use inverters to convert DC to AC. Because the largest practical inverter size to date is about 500 kVA, designers are building 1000 kVA transformers by placing two inverter-connected windings in one transformer. In this way the transformer has to have two separate windings to accept completely separate inputs. Design issues also stem from running cables long distances to convert from DC to AC.

1.20 Other Aspects

What are the standard connections of the PV application transformers? Some profess a special three-winding configuration, where the secondary consists of dual windings connected in a special manner to take care of the extreme voltage variations encountered by DPV-GTs.

Shielding requirements include electrostatic shielding, protective shielding, and harmonic filtration shielding.

Inverter technology has been slow to advance, because it is an electronic technology. It remains to be seen whether this comparative disadvantage will be a fatal flaw in the advancement of solar technology to the same level as wind farms in the renewable energy arena.

The size of the solar farm is limited by inverter technology, because inverters can currently only be built to about 500 kVA. This means that nearly all solar applications use pairs of 500 kVA inverters to drive the transformer and produce about 1000 kVA. Increasing the size by adding more inverters into one transformer box is extremely difficult due to complexities associated with the size of the box required and the practicalities of running cabling to convert from DC to AC.

Some core needs of a DPV-GTs are

- Efficient heat management: The heat generated due to uneven cooling of the coils leads to the creation of hot spots. This leads to premature breakdown of the transformers.
- Lower harmonics and grid disturbances.
- The ability to withstand harsh weather conditions, temperature, seismic levels, etc.

If necessary, DPV-GTs are designed and constructed to meet and exceed earthquake standards. Sometimes DPV-GTs are rated for installation in the highest earthquake rating zones. In addition, they can incorporate a variety of fluids, including less flammable fluids required for enclosed applications.

A DPV grid step-up transformer is especially designed to meet the solar industry's need for reliable service in remote locations, and it should offer advanced fault survivability/capability.

1.21 Conclusions

A DPV-GT solar converter step-up transformer is uniquely designed to connect solar farms to the electricity grid at large-scale solar power installations. These power-generating capacities are slowly increasing in the voltage levels and volt-amp levels. For example, in present-day applications solar farms have gone up to 69 kV and as high as 10 MVA capacity.

Reliable and efficient, step-up transformers are engineered solutions with the necessary design flexibility needed for the solar industry. The DPV-GT is designed for the additional loading associated with nonsinusoidal harmonic frequencies often found in inverter-driven transformers, and the design with multiple windings will be considered in case it can reduce transformer cost, minimize a transformer footprint, and provide required functionality. The shell-type transformers can also be considered for this application.

The duty cycle seen in solar farms may not be as severe as that seen in wind farms, but solar power has its share of special considerations that affect transformer design. Those engaged in harnessing solar energy need to pay heed to these special needs to ensure that the solar installation is cost effective and reliable.

Bibliography

1. Considerations for power transformers applied in distributed photovoltaic (DPV)-grid application, DPV-grid transformer task force members, Power Transformers Subcommittee, IEEE-TC, Hemchandra M. Shertukde, Chair, Mathieu Sauzay, Vice Chair, Aleksandr Levin, Secretary, Enrique Betancourt, C. J. Kalra, Sanjib K. Som, Jane Verner, Subhash Tuli, Kiran Vedante, Steve Schroeder, Bill Chu, white paper in preparation for final presentation at the IEEE-TC conference in San Diego, CA, April 10–14, 2011.
2. C57.91: IEEE guide for loading mineral-oil-immersed transformers, 1995, Correction 1-2002.
3. C57.18.10a: IEEE standard practices requirements for semiconductor power rectifier transformers, 1998, amended in 2008.

4. C57.110: IEEE recommended practice for establishing liquid-filled and dry-type power and distribution transformer capability when supplying non-sinusoidal load current, 2008.
5. C57.116: IEEE guide for transformers directly connected to generators, 1989.
6. C57.129: IEEE standard for general requirements and test code for oil-immersed HVDC convertor transformer, 1999 (2007, approved).
7. Standard 1547.4: Draft guide for design, operation and integration of distributed resource island systems with electric power system (only 1547.1 is there), 2005.
8. UL 1741: A safety standard for distributed generation, 2004.
9. Buckmaster, David, Hopkinson, Phil, Shertukde, Hemchandra, Transformers used with alternative energy sources—Wind and solar, Technical presentation, April 11, 2011.
10. Standard 519: Recommended practices and requirements for harmonic control in electrical power systems, 1992.
11. IEEE P1433: A standard glossary of power quality terminology, 1999.
12. DISPOWER project (Contract No. ENK5-CT-2001-00522), Identification of general safety problems, definition of test procedures and design-measures for protection, 2004.
13. DISPOWER project (Contract No. ENK5-CT-2001-00522), Summary report on impact of power generators distributed in low voltage grid segments, 2005 (http://www.pvupscale.org).
14. IEA PVPS Task V, report IEA-PVPS T5-01: 1998, Utility aspects of grid connected photovoltaic power systems.
15. IEC 61000-3-2: 2005, EMC—Part 3-2: Limits—Limits for harmonic current emissions equipment input current up to and including 16 A per phase.
16. IEEE 929: 2000, Recommended practice for utility interface of photovoltaic (PV) systems.
17. Engineering Recommendation G77/1: 2000, Connection of single-phase inverter connected photovoltaic (PV) generating equipment of up to 5 kW in parallel with a distribution network operators (DNO) distribution system.
18. IEC/TS 61000-3-4, EMC—Part 3-4: Limits—Limitation of emission of harmonic currents in low-voltage power supply systems for equipment with rated current greater than 16 A, October 30, 1998.
19. IEC/TR3 61000-3-6, EMC—Part 3-6: Limits—Assessment of emission limits for distorting loads in MV and HV power systems—Basic EMC publication, 1996-10.
20. Cobben, J. F., Heskes, P. J., Moor de H. H., Harmonic distortion in residential areas due to large scale PV implementation is predictable. *DER-Journal*, January 2005.
21. Cobben, J. F., Kling, W. L., Heskes, P. J., Oldenkamp, H., Predict the level of harmonic distortion due to dispersed generation, 18th International Conference on Electricity Distribution (CIRED), Turin, Italy, June 2005.
22. Cobben, J. F., Kling, W. L., Myrzik, J. M., Making and purpose of harmonic fingerprints, 19th International Conference on Electricity Distribution (CIRED), Vienna, Austria, May 2007.
23. Oldenkamp, H., De Jong, I., Heskes, P. J. M., Rooij, P. M., De Moor, H. H. C., Additional requirements for PV inverters necessary to maintain utility grid quality in case of high penetration of PV generators, 19th EC PVSEC, Paris, France, 2004, pp. 3133–3136.
24. Cobben, J. F. G., Power quality implications at the point of connection, Dissertation, University of Technology Eindhoven, 2007.

25. IEA-PVPS Task V, Report IEA-PVPS T5-2: 1999, Demonstration test results for grid interconnected photovoltaic power systems.
26. Halcrow Gilbert Associates, Department of Trade and Industry, Coordinated experimental research into power interaction with the supply network—Phase 1 (ETSU S/P2/00233/REP), 1999 (http://www.dti.gov.uk/publications).
27. UNIVERSOL project (Contract No. NNE5-293-2001), quality impact of the photovoltaic generator "Association Soleil-Marguerite" on the public distribution network, EDF-R&D, 2004.
28. IEC 61000-4-7: 2002, Electromagnetic compatibility (EMC)—Part 4-7: Testing and measurement techniques—General guide on harmonics and interharmonics measurements and instrumentation, for power supply systems and equipment connected thereto.
29. IEC 61000-2-12: 2003. Electromagnetic compatibility (EMC)—Part 2-12: Environment—compatibility levels for low-frequency conducted disturbances and signaling in public medium voltage power supply systems.
30. Hong, Soonwook, Zuercher–Martinson, Michael, Harmonics and noise in photovoltaic (PV) inverter and the mitigation strategies, white paper, Solectria, Lawrence, MA.
31. Harmonic analysis report, multiple loads, Allied Industrial Marketing, Cedarsburg, WI. September 2011.

2

Use of Distributed Photovoltaic Grid Power Transformers

As the name suggests, distributed photovoltaic grid transformers (DPV-GTs) are primarily used to transmit solar energy harnessed using photovoltaic (PV) systems, where a DC energy source is suitably converted by inverters to the AC form. The initial ratings (up to 1000 VA, 1.1 kV) of these transformers were suitable for residential application. Over the course of the last decades the suitability has increased to a level now of up to 10,000 VA, 33 kV. In the middle eastern and western parts of the United States, this application has found increased applications in industrial parks and also in solar farms that have harnessed energy up to 69 KV, 35 MVA, as in California. Many more solar farms are now creeping up in Arizona, Colorado, Oregon, and Nevada. Around the world, solar farms find extensive applications in Europe, Asia, and Africa. In Europe, Germany has been a leader in this field. In Asia, India and China have embraced harnessing solar energy, and thus transformer manufacture has increased in the small and medium transformer ranges.

2.1 DPV-GT Solar Converter Step-Up Transformers

The DPV-GT solar converter step-up transformer is uniquely designed to connect solar farms to the electricity grid at large-scale solar power installations.

Reliable and efficient step-up transformers are engineered solutions with the necessary design flexibility needed for the solar industry. The DPV-GT is designed for the additional loading associated with nonsinusoidal harmonic frequencies often found in inverter-driven transformers, and it has an innovative system of multiple windings that reduce transformer costs and minimize a transformer footprint.

DPV-GTs are designed and constructed to meet and exceed earthquake standards. The DPV-GT is rated for installation in the highest earthquake rating zones. In addition, it incorporates a variety of fluids, including less flammable fluids required for enclosed applications. The DPV-GT features circular windings that spread the radial forces evenly over their circumference and have cooling ducts throughout the coils, eliminating hot spots that lead to premature breakdown and ultimately to transformer failure. The coil end blocking with heavy-duty three-gauge steel bracing and pressure plates

FIGURE 2.1
(See color insert) Typical power transformer used in a solar inverter type application. (Courtesy of Power Distribution, Inc. dba Onyx Power.)

takes on the axial forces exerted during a fault condition. These forces can cause telescoping of the coils, shortening transformer life.

The DPV-GT features an innovative design that includes round coils, a cruciform, mitered core with heavy-duty clamping, and a proprietary pressure plate design, as well as a premium no-load tap changer. The DPV grid step-up transformer is especially designed to meet the solar industry's need for reliable service in remote locations, and it offers superb fault survivability. These transformers also have multiple windings to cater to the lower kVA ratings of the inverters available in the market. The secondaries are generally in a single-story or double-story format (see Figure 2.1).

Depending on harmonic content, the user may wonder how to take it into account to define the kVA rating of the transformer. C57.129 and C57.18.10 use a definition based only on the fundamental frequency and propose ways to take additional losses due to harmonics into account during heat-run tests.

Transformers are critical components in solar energy production and distribution. Historically, transformers have "stepped-up" or "stepped-down" energy from nonrenewable sources. There are different types of solar transformers including distribution, station, substation, pad mounted, and grounding. All transformers have specialized needs that impact costs.

Solar power applications experience steady-state loading during inverter operation. When the sun is out, there is a dampened reaction process and more constant loading on the transformer.

Also, fault ride-through has not been defined for photovoltaic systems. This may be because it is easier to turn solar systems on and off quickly, or because regulatory requirements have not caught up with the young technology. This may change in the future, but so far there are no solar systems with such a requirement.

Harmonics in the solar inverter's typical harmonic content is below 1%, which has almost no impact on the system. The lower harmonic profile is because there are no generators and switching and protective controls such as those found on wind turbines. Solar transformers do require step-up duty. Yet, the inverter converts DC input from the PV array to AC voltage for the transformer in a smooth transition with no over-voltage from an unloaded circuit. Because solar transformers operate at a steady voltage, with the rated voltage controlled by inverters, voltage and load fluctuations are considerably lower than in wind turbines. Solar systems also operate close to their rated loads.

Solar power systems also have special design issues. The largest inverter size is about 500 kVA, designers are building 1000 kVA transformers by placing two inverter connected windings in one box as shown in Figure 2.2a. The transformer must have separate windings to accept completely separate inputs. Design issues also stem from running cables long distances to convert from DC to AC.

The "special" characteristic of this transformer is that it is designed to meet CEC efficiency of 99% (as requested by the customer). CEC efficiency compliance is a common requirement for transformers used in the solar power industry. The nameplate (see Figure 2.2b) below shows the other guaranteed details for this solar transformer including the winding connections.

Restrictions on inverter size also limit the size of solar systems. Increasing the size by adding more inverters into one transformer box is extremely difficult. With the required box size and running cabling to convert DC to AC, things get complex. The individual sizes of such units have grown to 1 MVA.

The key to solar transformers is to understand the variables in every system. Transformers need to customize to work with each particular system. So far, inverter technology has been slow to advance, and it remains to be seen whether this comparative disadvantage will be a fatal flaw in the advancement of solar technology to the same level as wind farms.

Facing ever-increasing worldwide energy demand, the reliable and environmentally friendly use of natural energy sources is one of the biggest challenges of our time. Alongside wind and water, the sun—clean, CO_2-neutral, and limitless—is our most valuable resource. In order to make renewable the dominant energy source all over the world, everyone should aim to make them as affordable as conventional sources of energy. By combining innovations in renewable power generation with the smart grid and high-voltage transmission technology, we are able to be even more cost efficient as well as energy efficient. Several large companies like Siemens offer proven components along the entire solar power value chain. Such transformers, whether liquid-filled or GEAFOL cast-resin distribution transformers or power transformers, have been in service all around the world for decades. Such reliable and established technology is customized for state-of-the-art energy production.

(a)

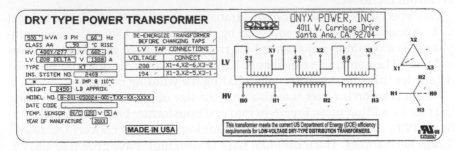

(b)

FIGURE 2.2
(See color insert) (a) DPV-GT dry-type (fiberglass insulated) for solar application. (Courtesy Power Distribution, Inc. dba Onyx Power.) Note the simple winding configuration on the secondary of the delta/wye neutral grounded configuration in a frontal view. (b) Nameplate details of DPV-GT dry-type (fiberglass insulated) for solar application for transformer in Figure 2.2a. (Courtesy of Power Distribution, Inc. dba Onyx Power.)

2.2 Transformers for Solar Power Solutions

2.2.1 Photovoltaic Power Plants

Photovoltaic (PV) systems use solar cells bundled in solar panels to produce DC current. Depending on the design of the photovoltaics plant, several panels are connected to a rectifier to convert the produced DC current into

AC current. In the next step, distribution or static converter transformers (GEAFOL or liquid immersed) transmit the energy to medium voltage level up to 36 kV. Then it is bundled and a medium power transformer steps it further up to a high-voltage level.

2.2.2 Concentrated Solar Power

Concentrated solar power (CSP) uses lenses or mirrors to bundle the sunlight and concentrate it on a small spot. The concentrated heat runs a steam turbine connected to a generator (thermo-electricity). Usually the turbine generates a higher power level than photovoltaics do, so a medium power transformer is sufficient to connect the CSP plant to the grid.

2.2.3 PV Distribution Transformers

Step-up transformers connect photovoltaic plants to the grid. As the conditions in solar power plants are rather severe, those transformers need to withstand high temperatures and harsh weather conditions. Sizing of these transformers is a crucial factor when planning a PV power plant, as too large rated power can lead to instabilities and economic disadvantages as well as too small transformer power might not exploit the full capability of the plant erected. Solar inverters or PV inverters for photovoltaic systems transform DC power generated from the solar modules into AC power and feed this power into the network. A special multiple winding design of the transformer enables several PV panel strings to be connected to the grid with a minor number of transformers. Examples are shown in Figures 2.3 and 2.4 respectively.

FIGURE 2.3
(See color insert) Many transformers are pad-mount types.

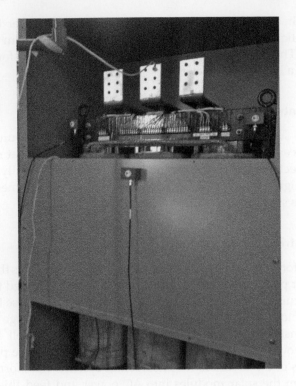

FIGURE 2.4
(See color insert) Dry-type CSP power transformers in solar PV grid application. (Courtesy of Diagnostic Devices Inc.)

2.3 Concentrated Solar Power (CSP) Transformers

Transformers in CSP plants usually belong to the group of medium power transformers. As a CSP generates power by driving a steam turbine, the duty for the transformer is very close to its common task of stepping up generated power in conventional power plants.

2.4 Medium Power Transformers

Electricity generated by solar power plants has to be transmitted to the areas of consumption. Therefore, medium power transformers increase the voltage level of the generated electricity to usually about 110 kV or 220 kV to bring forward the bundled energy efficiently. Power ranges up to 200 MVA or even higher can come in several variations: with an offload tapping switch or on-load tap changers, with a combination of the two, or with reconnect devices under the cover or in the reconnect dome. The range of possibilities includes separate winding transformers and autotransformers, as well as three-phase and single-phase designs. The precise requirements vary from device to device and from site to site. That's why each transformer must be almost as unique as a fingerprint when it comes to voltage, power, climate efficiency, network topology, permissible noise level, and other factors.

2.5 Concentrated Photovoltaic (CPV) Systems Transformers

In these systems, solar energy is concentrated with the use of parabolic solar panels. The solar energy is concentrated at the focal point of these panels, thus increasing the efficiency of converting the solar energy into electrical energy. The efficiency of such systems is reported to have climbed to 41% and in some cases to 65% with the use of maximum power point tracking (MPPT) systems. A present debate is brewing comparing the success and efficiency of the CPV versus CSP systems. Over time the CPV system seems to present a very profitable system for the future.

A maximum power point tracker (or MPPT) is a high-efficiency DC-to-DC converter that functions as an optimal electrical load for a photovoltaic (PV) cell, most commonly for a solar panel or array, and converts the power to a voltage or current level that is more suitable to whatever load the system is designed to drive. PV cells have a single operating point where the values of the current (I) and voltage (V) of the cell result in a maximum power output. These values correspond to a particular resistance, which is equal to V/I as specified by Ohm's law. A PV cell has an exponential relationship between current and voltage, and the maximum power point (MPP) occurs at the knee of the curve, where the resistance is equal to the negative of the differential resistance ($V/I = -dV/dI$). Maximum power point trackers utilize some type of control circuit or logic to search for this point and thus to allow the converter circuit to extract the maximum power available from a cell. Traditional solar inverters perform MPPT for an entire array as a whole. In such systems the same current, dictated by the inverter, flows through all panels in the string. But because different panels have different IV curves (i.e., different MPPs) (due to manufacturing tolerance, partial shading, etc.), this architecture means some panels will be performing below their MPP, resulting in a loss of energy.

A typical CPV system is shown in Figure 2.5 and can be similarly represented by a single line diagram as in Figure 1.1 including the DPV-GT. Compared to nonconcentrated photovoltaics, CPV systems can save money on the cost of the solar cells, because a smaller area of photovoltaic material is required. Because a smaller PV area is required, CPVs can use the more

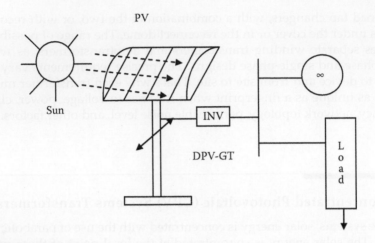

FIGURE 2.5
Swiveling solar panel of a CPV system in the direction of the arrow with the sun's rays concentrated with mirrors.

expensive high-efficiency tandem solar cells. To get the sunlight focused on the small PV area, CPV systems require spending extra money on concentrating optics (lenses or mirrors), solar trackers, and cooling systems. Because of these extra costs, CPV is far less common today than nonconcentrated photovoltaics. However, ongoing research and development is trying to improve CPV technology and lower costs.

Bibliography

1. Basso, Thomas S., High-penetration, grid-connected photovoltaic technology codes and standards, 33rd IEEE Photovoltaic Specialists Conference, 2008, pp. 1–4.
2. Weidong, Yang, Xia, Zhou, Feng, Xue, Impacts of large scale and high voltage level photovoltaic penetration on the security and stability of power system, Asia-Pacific Power and Energy Engineering Conference, 2010, pp. 1–5.
3. Seo, H. C., Kim, C. H., Yoon, Y. M., Jung, C. S., Dynamics of grid-connected photovoltaic system at fault conditions, Asia and Pacific Transmission and Distribution Conference and Exposition, 2009, pp. 1–4.
4. Phuttapatimok, S., Sangswang, A., Seapan, M., Chenvidhya, D., Kirtikara, K., Evaluation of fault contribution in the presence of PV grid-connected systems, 33rd IEEE Photovoltaic Specialists Conference, 2008, pp. 1–5.
5. Dash, Prajna Paramita, Kazerani, Mehrdad, Study of islanding behavior of a grid connected photovoltaic system equipped with a feed-forward control scheme, 36th Annual Conference on IEEE Industrial Electronics Society, November 2010, pp. 3228–3234.

6. Wang, Li, Lin, Ying-Hao, Dynamic stability analyses of a photovoltaic array connected to a large utility grid, IEEE Power Engineering Society Winter Meeting, January 2000, vol. 1, pp. 476–480.
7. Rodriguez, C., Amaratunga, G. A. J., Dynamic stability of grid-connected photovoltaic systems, IEEE Power Engineering Society General Meeting, June 2004, pp. 2193–2199.
8. Wang, Li, Lin, Tzu-Ching, Dynamic stability and transient responses of multiple grid connected PV systems, IEEE/PES Transmission and Distribution Conference and Exposition, 2008, pp. 1–6.
9. Yazdani, Amirnaser Dash, Prajna Paramita, A control methodology and characterization of dynamics for a photovoltaic (PV) system interfaced with a distribution network, *IEEE Trans. on Power Delivery*, 2009, vol. 24, pp. 1538–1551.
10. Edrington, Chris S., Balathandayuthapani, Saritha, Cao, Jianwu, Analysis and control of a multi-string photovoltaic (PV) system interfaced with a utility grid, IEEE Power and Energy Society General Meeting, 2010, pp. 1–6.
11. Edrington, Chris S., Balathandayuthapani, Saritha, and Cao, Jianwu, Analysis of integrated storage and grid interfaced photovoltaic system via nine-switch three-level inverter, IECON 2010-36th Annual Conference on IEEE Industrial Electronics Society, November 2010, pp. 3258–3262.
12. Ito, T., Miyata, H., Taniguchi, M., Aihara, T., Uchiyama, N., Konishi, H., Harmonic current reduction control for grid-connected PV generation systems, International Power Electronics Conference, 2010, pp. 1695–1700.
13. Hojo, Masahide, Ohnishi, Tokuo, Adjustable harmonic mitigation for grid-connected photovoltaic system utilizing surplus capacity of utility interactive inverter, 37th IEEE Power Electronics Specialists Conference, 2006, pp. 1–6.
14. Hossein Hosseini, Seyed, Sarhangzadeh, Mitra, Sharifian, Mohammad B. B., Sedaghati, Farzad, Using PV in distribution network to supply local loads and power quality enhancement, International Conference on Electrical and Electronics Engineering, 2009, pp. 249–253.
15. Chen, Xiaogao, Fu, Qing, Wang, Donghai, Performance analysis of PV grid connected power conditioning system with UPS, 4th IEEE Conference on Industrial Electronics and Applications, 2009, pp. 2172–2176.
16. Li, Jing, Zhuo, Fang, Liu, Jinjun, Wang, Xianwei, Wen, Bo, Wang, Lin, Ni, Song, Study on unified control of grid-connected generation and harmonic compensation in dual-stage high-capacity PV system, IEEE Energy Conversion Congress and Exposition, 2009, pp. 3336–3342.
17. Chen, Xiaogao, Fu, Qing, Infield, David, Yu, Shijie, Modeling and control of Zsource grid-connected PV system with APF function, 44th International Universities Power Engineering Conference, 2009, pp. 1–5.
18. Li, Hongyu, Zhuo, Fang, Wang, Zhaoan, Lei, Wanjun, Wu, Longhui, A novel time-domain current-detection algorithm for shunt active power filters, *IEEE Trans. on Power Systems*, 2005, vol. 20, pp. 644–651.
19. Kim, Gyeong-Hun, Seo, Hyo-Rong, Jang, Seong-Jae, Park, Sang-Soo, Kim, Sang-Yong, Performance analysis of the anti-islanding function of a PV-AF system under multiple PV system connections, International Conference on Electrical Machines and Systems, 2009, pp. 1–5.
20. Bhattacharya, Indranil, Deng, Yuhang, Foo, Simon Y., Active filters for harmonics elimination in solar photovoltaic grid-connected and stand-alone systems, 2nd Asia Symposium on Quality Electronic Design, 2010, pp. 280–284.

21. Walker, Geoffrey R., Sernia, Paul C., Cascaded DC-DC converter connection of photovoltaic modules, IEEE Trans. on Power Electronics, July 2004, vol. 19, pp. 1130–1139.
22. Campbell, Ryan C., A circuit-based photovoltaic array model for power system studies, 39th North American Power Symposium, September 2007, pp. 97–101.
23. dSPACE Inc, dSPACE™ 1103 V6.5 Manual, 2007.
24. Esram, Trishan, Chapman, Patrick L., Comparison of photovoltaic array maximum power point tracking techniques, IEEE Trans. on Energy Conversion, June 2007, vol. 22, pp. 439–449.
25. Xiao, Weidong, Ozog, Nathan, Dunford, William G., Topology study of photovoltaic interface for maximum power point tracking, IEEE Trans. on Industrial Electronics, June 2007, vol. 54, pp. 1696–1704.
26. Chung, Se-Kyo, A phase tracking system for three phase utility interface inverters, *IEEE Trans. on Power Electronics*, May 2000, vol. 15, pp. 431–438.
27. IEEE Standard 519-1992: IEEE recommended practices and requirements for harmonic control in electrical power systems, p. 78.

3

Voltage Flicker and Variation in Distributed Photovoltaic Grid Transformers

IEEE Standard 519 clearly states the requirements for harmonic control in transformers connected to grids which may contribute to the quality of the signal. This is connected to several nth harmonics that may cause a change in the nominal value of the voltage at the terminals of the distributed photovoltaic grid transformer (DPV-GT). Besides, the load connected at the grid on the AC end may cause a nonuniform time requirement as in the case of battery charging scheduled by customers for their electrical vehicles.

Generally in a traditional solar power system, voltage is generated at a low voltage (28 volts DC) and transmitted by a step-up transformer at a higher voltage, say 33 kV, to a load point where it is available for use at a lower voltage, say 115 volts AC. In some cases this energy is tied to the grid. The supply voltage at the delivery point is expected to be within ±10% of the nominal value. Stricter limits may apply locally as well as nationally. Voltage variation in distribution systems is due to the variation of loads, for example, during nights when temperatures drop or during summer days when temperatures rise. This is averted at the MV levels by the use of automatic load tap-changers. At the load points this is achieved by use of off-circuit tap changers that can vary load in steps of 2.5% of the nominal voltage.

3.1 Flicker

Fluctuations in frequency that range from 1 to 20 Hz can cause light flicker that may affect electronic equipment that relies on constant power. Short-term flicker value P_{st} is calculated according to a process over a predefined observation interval. Long-term flicker P_{lt} is calculated as the cubic average of several P_{st} values. In the standard IEC 61000, the observation intervals and the limiting values for P_{st} and P_{lt} are specified as follows:

1. Any dramatic change in current will cause a dramatic change in voltage. This can occur whenever the generator main breaker opens, or during clouding.
2. Flicker is limited to 2 volts (2.5%) on a 120V base in urban areas or 5 volts (4.17%) on a 120V base in rural areas.

3.2 Voltage Fluctuations

Voltage fluctuations are due to fluctuations in load current. Examples of load that lead to voltage fluctuations are furnaces, copy machines, and refrigerators. Renewable sources of energy that show a fast fluctuation in output power (e.g., wind power, solar power) are also potential sources of voltage fluctuation. Central Electricity Authority (CEA) regulations on grid connectivity specify that the voltage fluctuation limit for step changes that may occur repetitively is 1.5%, and for occasional fluctuations other than step changes, the maximum permissible limit proposed is 3%.

Alternately, the output of the DPV-GT grid-tied generation system fluctuates with the hourly changes in solar radiation with the movement of the clouds on a clear day or a cloudy day. This may cause a fluctuation in the solar power flow and thus the related voltage fluctuation. This can be more predominant than voltage fluctuation caused by varying loads in a concentrated PV distribution system. It is critical to closely control and monitor these changes, as it is likely under certain circumstances as evaluated by the over-/under-voltage (OVP/UVP) and over-/under-frequency (OFP/UFP) schemes. Grid-tied DPV-GT systems need to have OVP/UVP and/or OFP/UFP detection systems that will cause the PV inverter to stop supplying power to the DPV-GT and the grid if the frequency or the amplitude of the voltage at the point of common coupling (PCC) between the customer (load) and the utility varies beyond prescribed limits. These serve to protect customers' equipment and also serve as anti-islanding detection methods. The OVP/UVP relaying systems play an important part in deciding when to disconnect or avoid disconnecting from the grid (as described in detail in Chapter 6) by the passive and active methods used with the nondetection zone (NDZ) method.

Modern PV inverters used with DPV-GT configurations with multiple secondary windings and multiple inverters have the ability to control voltage and/or VARs at the point of common coupling (PCC) with the electric utility. Due to the VOLT/VAR control of the inverter, the PV station can actually improve voltage control on the feeder. In some cases, it might actually help to avoid voltage collapse in the event of severe system disturbances. The OVP/UVP play an important role in this monitoring and control through the supervisory control and data acquisition (SCADA) system in place.

Presently both the PV station distance from the substation and the DPV-GT and PV circuit penetration play critical roles in feeder performance (see Figure 3.1). Ohm's law considerations and kind of load (RLC) govern this pattern. The 100% peak PV ratio curve is almost linear, while the other curves are nonlinear.

Voltage flicker can be resolved by the use of high-speed insulated-gate bipolar transistor (IGBT) inverter-based voltage regulators that are always connected as shunt voltage regulators. An inverter-based VAR source can modulate the current flowing supply almost instantaneously and thus create

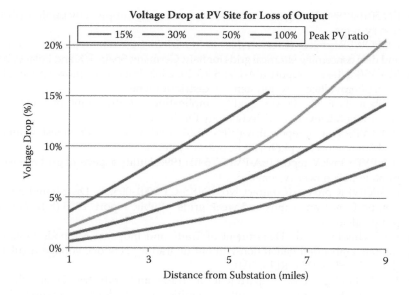

FIGURE 3.1
(See color insert) Percentage of voltage drops as a function of distance from substation and DPV-GT.

a correction voltage on the supply impedance. Additionally, the inverter will update the current many times per cycle to provide an effective control. Thus, a small variation in voltage typically signified as flicker can cause a significant change in lighting. This is defined by a measure called P_{st}. A value of $P_{st} = 1$ is tolerable, but any values higher than 1 can cause irritation in the human population and are undesirable. A P_{st} of 1 corresponds to 0.5% variation in voltage at a most sensitive region of frequency of 5 Hz. This IGBT scheme helps to optimize the value of P_{st} by managing the VARs and thus also limits the impact on protection system fault currents.

Bibliography

1. National Electrical Code (NEC) (NFPA 70-2005).
2. The Public Utilities Regulatory Act of 1978 (PURPA).
3. IEEE recommended practice for utility interface of photovoltaic (PV) systems (IEEE Standard 929-2000).
4. Underwriters Laboratories' (UL) testing standard UL 1741.
5. IEEE Standard 929.
6. IEEE Standard 1547.
7. Walton, S. J., Technical bulletin: Voltage regulation and flicker control using Vectek IGBT static VAR compensators, February 2005, Omniverter Inc., Ontario, Canada.

8. EN 50160:1999, Voltage characteristics of electricity supplied by public distribution systems.
9. DISPOWER project (Contract No. ENK5-CT-2001-00522), Appendix—Structure and data concerning electrical grids for Italy, Germany, Spain, UK and Poland, 2004.
10. IEA-PVPS Task V, report IEA-PVPS T5-10: 2002, Impacts of power penetration from photovoltaic power systems in distribution networks.
11. Cobben, J. F. G., Power quality implications at the point of connection, Dissertation University of Technology Eindhoven, 2007.
12. IEA-PVPS Task V, report IEA-PVPS T5-02: 1999, Demonstration test results for grid interconnected photovoltaic power systems.
13. IEA PVPS Task V, report IEA-PVPS T5-01: 1998, Utility aspects of grid connected photovoltaic power systems.
14. DISPOWER project (Contract No. ENK5-CT-2001-00522), Distributed generation on European islands and weak grids—Public report, 2005 (http://www.pvupscale.org).
15. EA Technology Ltd., Department of Trade and Industry, Methods to accommodate embedded generation without degrading network voltage regulation (ETSU K/EL/00230/REP), 2001.
16. EA Technology Ltd., Department of Trade and Industry, Likely changes to network design as a result of significant embedded generation (ETSU K/EL/00231/REP), 2001.
17. UMIST, ECONNECT, Department of Trade and Industry, Integration of operation of embedded generation and distribution networks (K/EL/00262/REP), 2002.
18. Halcrow Gilbert Associates Ltd., Department of Trade and Industry, Microgeneration network connection (renewables) (K/EL/00281/00/00), 2003.
19. IPSA Power, Smith Rea Energy, Department of Trade and Industry, Technical solutions to enable generation growth (K/EL/00278/00/0), 2003 (http://www.dti.gov.uk/publications).
20. Halcrow Gilbert Associates Ltd., Department of Trade and Industry, Co-ordinated experimental research into power interaction with the supply network—Phase 1 (ETSU S/P2/00233/REP), 1999.
21. Andrieu, C., Tran, T., The connection of decentralised energy producers to the low voltage grid (Le raccordement en basse tension des producteurs décentralises d'énergie), INPG/IDEA, 2003.
22. Kawasaki, N., Oozeki, T., Otani, K., Kurokawa, K., An evaluation method of the fluctuation characteristics of photovoltaic systems by using frequency analysis, *Solar Energy Materials and Solar Cells* 90 (2006), 3356–3363.
23. Paatero, J. V., Lund, P. D., Effects of large-scale photovoltaic power integration on electricity distribution networks, *Renewable Energy* 32 (2007), 216–234.
24. Ueda, Y., et al., Analytical results of output restriction due to the voltage increasing of power distribution line in grid-connected clustered PV systems, 31st IEEE Photovoltaic Specialists Conference Proceedings (2005), pp. 1631–1634.
25. Takeda, Y., et al., Test and study of utility interface and control problems for residential PV systems in Rokko Island 200kW test facility, 20th IEEE Photovoltaic Specialists Conference (1988), pp. 1062–1067.

4

Harmonics and Waveform Distortion (Losses, Power Rating) in Distributed Photovoltaic Grid Transformers

Grid-tied distributed photovoltaic grid transformer (DPV-GT) systems provide a voltage supply that is never a pure sine wave. DPV-GT systems are subjected to disturbances in the form of harmonics and interharmonics that are superimposed on the supply voltage. This causes the power quality phenomena which can be caused both by voltage and current harmonics. The harmonics are defined as multiples of frequencies of the fundamental component (i.e., 60 or 50 Hz) and are defined by their spectral distribution and values in the frequency range as defined by IEEE P1-433-A. In addition, there are interharmonics that are voltage or current harmonics with frequencies not as multiples of the fundamental frequency but as discrete frequencies or as a part of broadband frequency as defined by IEC 61000-2-1.

DPV-GTs are subjected to waveform distortions due to harmonics that are induced into the system by the inverter mechanism. The solar inverter system's typical harmonic content is less than 1%, which has almost no impact on the system. The lower harmonic profile is because there are no generators and switching and protective controls such as those found on wind turbines. The other harmonics introduced in the system which affect the DPV-GTs is mainly due to switching caused by the system generating intermittent energy due to the clouds that may affect the amount of sun's rays falling on the PV panels and the time duration when it happens. There are harmonics induced into the system by nonlinear loads. The harmonic voltages have their origin in the harmonic currents drawn by these loads. Some examples of sources of harmonic currents are switch-mode power supplies, gas-discharge and fluorescent lamps, variable speed drives, uninterruptible power supplies, cyclo-converters, phase-angle controlled loads, arc furnaces, static-VAR compensators, and transformers. In addition, linear loads like resistors, capacitors, and inductors can be subjected to harmonic currents due to voltage distortion. These total harmonic levels could be as high as 100% for single-phase loads, but harmonic voltage distortion of greater than 8.5% is generally unlikely. The present-day design of transformer-less inverters also introduce even powered harmonics. PWM controlled inverters are examples of such devices. Even order harmonics are also caused by loads that have asymmetrical i-v characteristics.

The interharmonics are created by amplitude modulation of load current due to static converter switching of semiconductors in the inverter circuits, which is not synchronized with the system frequency or when the DPV-GTs are subject to magnetic circuit saturation. Typical sources of interharmonics are cyclo-converters, static frequency converters, arc welders, arc furnaces, induction motors, wind turbine generators, low-frequency power-line communication carriers, and loads controlled by integral cycle.

The negative effects of harmonics on transformers are commonly unnoticed and disregarded until an actual failure happens. Generally, transformers designed to operate at rated frequency have had their loads replaced with nonlinear types, which inject harmonic currents into the system. Consequently, transformers that have operated adequately for long periods have failed in a comparatively short time. Additional losses are introduced by harmonic contents of the signal, especially the eddy current losses and hysteresis losses. These are further accentuated by the nonlinear loads DPV-GTs are subjected to in their daily operations. For nonlinear loads, all standard DOE 10 CFR Part 431 distribution transformers have to be derated to allow for additional heat due to harmonic losses. Harmonic mitigating transformers are superior to K-rated and general-purpose transformers in that they reduce voltage distortion (flat-topping) and power losses due to current harmonics created by single-phase, nonlinear loads such as computer equipment. Secondary windings are arranged to cancel zero-sequence fluxes and eliminate primary winding circulating currents. They treat zero-sequence harmonics (3rd, 9th, and 15th) within the secondary windings and fifth and seventh harmonics upstream with appropriate phase shifting. Dual-output, phase-shifting harmonic mitigating transformers provide extremely low output voltage distortion and input current distortion even under severe nonlinear load conditions (data centers, Internet service providers, telecom sites, call centers, broadcasting studios, etc.). Combining zero-sequence flux cancellation with phase shifting treats 3rd, 5th, 7th, 9th, 15th, 17th, and 19th harmonics within its secondary windings.

In addition, harmonics and interharmonics have a wide range of effect on DPV-GT network components like thermal overheating, magnetic saturation, and system resonance. In turn these cause voltage flicker and fluctuations of the system and grid voltages.

4.1 Definition of Harmonics

Harmonics are AC voltages and currents with frequencies that are integer multiples of the fundamental frequency. On a 60-Hz system, this could include second-order harmonics (120 Hz), third-order harmonics (180 Hz), fourth-order harmonics (240 Hz), and so on. Normally, only odd-order

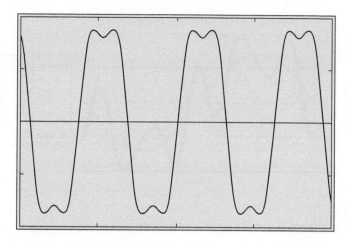

FIGURE 4.1
(See color insert) Fundamental frequency with an in-phase third-harmonic frequency.

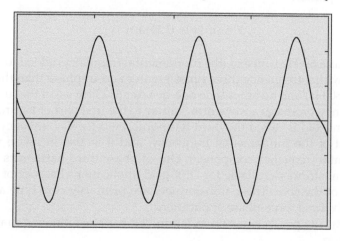

FIGURE 4.2
(See color insert) Fundamental frequency with an out-of-phase third-harmonic frequency.

harmonics (3rd, 5th, 7th, 9th) occur on a three-phase power system. If you observe even-order harmonics on a three-phase system, you more than likely have a defective rectifier in your system.

If you connect an oscilloscope to a 120 V receptacle, the image on the screen usually is not a perfect sine wave. It may be very close, but it will likely be different in one of several ways. It might be slightly flattened or dimpled as the magnitude approaches its positive and negative maximum values (Figure 4.1). Or perhaps the sine wave is narrowed near the extreme values, giving the waveform a peaky appearance (Figure 4.2). More than likely, random deviations from the perfect sinusoid occur at specific locations on the sine wave during every cycle (Figure 4.3).

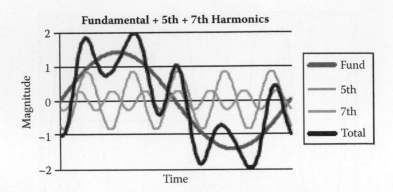

FIGURE 4.3
(See color insert) Fundamental and odd harmonics (fifth and seventh).

The flattened and dimpled sinusoid in Figure 4.1 has the following mathematical equation:

$$Y = \sin(x) + 0.25\sin(3x) \qquad (4.1)$$

This means a 60-Hz sinusoid (the fundamental frequency) added to a second sinusoid with a frequency three times greater and in-phase than the fundamental (180 Hz) and an amplitude one-quarter (0.25 times) of the fundamental frequency produces a waveform similar to the first part of Figure 4.1. The 180-Hz sinusoid is called the third harmonic, because its frequency is three times that of the fundamental frequency, and it is also in phase with the fundamental frequency component. Out-of-phase third harmonics can also exist in transformers suitable for DPV grid applications because of the construction of the core. These transformers are primarily core type and have three limbs for a three-phase application.

Similarly, the peaky sinusoid in Figure 4.2 has the following mathematical equation:

$$Y = \sin(x) - 0.25\sin(3x) \qquad (4.2)$$

This waveform has the same composition as the first waveform, except the third harmonic component is out of phase with the fundamental frequency, as indicated by the negative sign preceding the "0.25 sin (3x)" term. This subtle mathematical difference produces a very different appearance in the waveform.

The waveform in Figure 4.3 contains several other harmonics in addition to the third harmonic. Some are in phase with the fundamental frequency and others are out of phase. As the harmonic spectrum becomes richer in harmonics, the waveform takes on a more complex appearance, indicating more deviation from the ideal sinusoid as shown in Figure 4.4. A rich

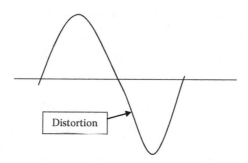

FIGURE 4.4
Slightly distorted fundamental frequency sinusoidal wave form as from Figure 4.3.

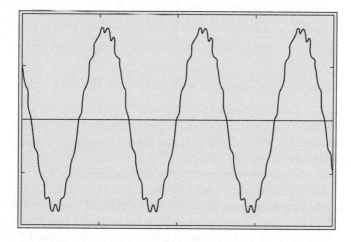

FIGURE 4.5
(See color insert) Distorted sinusoid containing all the harmonics.

harmonic spectrum may completely obscure the fundamental frequency sinusoid, making a sine wave unrecognizable.

Frequency analysis of harmonics when the magnitudes and orders of harmonics are known, reconstructing the distorted waveform is simple. Adding the harmonics together, point by point, produces the distorted waveform. The waveform in Figure 4.5 is synthesized in Figure 4.6 by adding the magnitudes of the two components, the fundamental frequency (red waveform), and the third harmonic (blue waveform) for each value of x, which results in the green waveform.

Decomposing a distorted waveform into its harmonic components is considerably more difficult. This process requires Fourier analysis, which involves a fair amount of calculus. The Fourier spectrum analysis provides the power embedded in each of the frequency components of the power signal being handled. However, electronic equipment has been developed to perform this

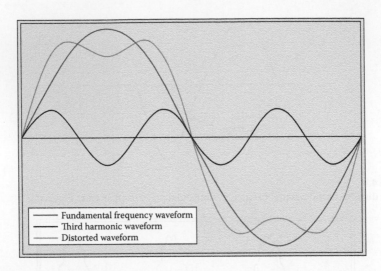

FIGURE 4.6
(See color insert) Synthesized waveforms with fundamental and third in-phase harmonic component.

analysis on a real-time basis. Some manufacturers offer a three-phase power analyzer that can digitally capture three-phase waveforms and perform a host of analysis functions, including Fourier analysis, to determine harmonic content. Other manufacturers offer similar capabilities for single-phase applications. Easy-to-use analyzers like these can help detect and diagnose harmonic-related problems on most power systems. Once these harmonics are ascertained, it is easy to use ready-made formulas to find the actual contents of the eddy current and hysteresis losses that occur in the DPV-GTs.

Harmonics are either current or voltage components of frequencies which are multiples of the fundamental frequency and are superimposed on the fundamental waveform. Although individually the harmonic components are sinusoidal, their superposition onto the fundamental waveform results in a distorted waveform as shown in Figure 4.3. The higher the harmonic content, the greater is the waveform distortion. Typical measured values of harmonic voltage distortion are in the range of 0% to 10% THD-v and typical current distortion ranges from 0% to more than 100% THD-i. When harmonic distortion is relatively high, the root-mean-square (RMS) value of circuit current and voltage can increase significantly. This can impose stress on electrical equipment and wiring and cause additional heating and reduced life expectancy of DPV-GTs.

Total voltage harmonic distortion (TVHD) of the voltage waveform is defined as the ratio of the root-sum-square value of the harmonic content of the voltage to the root-mean-square value of the fundamental voltage.

$$V_{TVHD} = \text{Sqrt}\left(V_2^2 + V_3^2 + V_4^2 + V_5^2 + \ldots\right)/V_1 \times 100\% \qquad (4.3)$$

Total current harmonic distortion (TCHD) of the current waveform is the ratio of the root-sum-square value of the harmonic content of the current to the root-mean-square value of the fundamental current:

$$I_{TCHD} = \text{Sqrt}\left(I_2^2 + I_3^2 + I_4^2 + I_5^2 + ...\right)/I_1 \times 100\% \qquad (4.4)$$

Total current demand distortion (TCDD) of the current waveform is the ratio of the root-sum-square value of the harmonic current to the maximum demand load current:

$$I_{TCDD} = \text{Sqrt}\left(I_2^2 + I_3^2 + I_4^2 + I_5^2 + ...\right)/I_l \times 100\% \qquad (4.5)$$

where I_l is the maximum demand load current over the past 12 months of operation of the system.

4.2 Factors That Cause Harmonics

If harmonic voltages are not generated intentionally, where do they come from? One common source of harmonics is iron core devices like transformers. The magnetic characteristics of iron are almost linear over a certain range of flux density but quickly saturate as the flux density increases.

This nonlinear magnetic characteristic is described by a hysteresis curve. Because of the nonlinear hysteresis curve, the excitation current waveform is not sinusoidal. A Fourier analysis of the excitation current waveform reveals a significant third harmonic component, making it similar to the waveform shown in Figure 4.2.

Core iron is not the only source of harmonics. Inverters produce some fifth harmonic voltages due to magnetic flux distortions that occur in the silicon circuitry used like IGBT and integrated gate commutated/controlled thyristors (IGCTs). Other producers of harmonics include nonlinear loads like rectifiers, adjustable-speed motor drives, welders, arc furnaces, voltage controllers, and frequency converters. The latter are typically associated with the load circuit.

Semiconductor switching devices produce significant harmonic voltages as they abruptly chop voltage waveforms during their transition between conducting and cutoff states. Inverter circuits are notorious for producing harmonics and are in widespread use today. An adjustable-speed motor drive is one application that makes use of inverter circuits, often using pulse width modulation (PWM) synthesis to produce the AC output voltage. Various synthesis methods produce different harmonic spectra. Regardless of the method used to produce an AC output voltage from a DC input voltage, harmonics will be present on both sides of the inverter and must often be mitigated.

4.3 Effects of Harmonics

Besides distorting the shape of the voltage and current sinusoids, what other effects do harmonics cause? Because harmonic voltages produce harmonic currents with frequencies considerably higher than the power system fundamental frequency, these currents encounter much higher impedances as they propagate through the power system than does the fundamental frequency current. This is due to the "skin effect," which is the tendency for higher-frequency currents to flow near the surface of the conductor. Since little of the high-frequency current penetrates far beneath the surface of the conductor, less cross-sectional area is used by the current. As the effective cross section of the conductor is reduced, the effective resistance of the conductor is increased. This is expressed in the following equation:

$$R = \rho l / A \qquad (4.6)$$

where R is the resistance of the conductor, ρ is the resistivity of the conductor material, l is the length of the conductor, and A is the cross-sectional area of the conductor. The higher resistance encountered by the harmonic currents will produce a significant heating of the conductor, because heat produced—or power lost—in a conductor is I^2R, where I is the current flowing through the conductor.

This increased heating effect is often noticed in two particular parts of the power system: neutral conductors and transformer windings. Harmonics with orders that are odd multiples of the number three (3rd, 9th, 15th, and so on) are particularly troublesome, since they behave like zero-sequence currents. These harmonics, called triplen harmonics, are additive due to their zero-sequence-like behavior. They flow in the system neutral and circulate in delta-connected transformer windings, generating excessive conductor heating in their wake.

4.4 Reducing the Effects of Harmonics

Because of the adverse effect of harmonics on power system components, the IEEE developed Standard 519-1992 to define recommended practices for harmonic control. This standard also stipulates the maximum allowable harmonic distortion allowed in the voltage and current waveforms on various types of systems.

Two approaches are available for mitigating the effects of excessive heating due to harmonics, and a combination of the two approaches is often implemented. One strategy is to reduce the magnitude of the harmonic waveforms, usually by filtering. The other method is to use system components that can handle the harmonics more effectively, such as finely stranded conductors and K-factor transformers.

Harmonic filters can be constructed by adding an inductance (L) in series with a power factor correction capacitor (C). The series L-C circuit can be tuned for a frequency close to that of the troublesome harmonic, which is often the fifth. By tuning the filter in this way, you can attenuate the unwanted harmonic.

Filtering is not the only means of reducing harmonics. The switching angles of an inverter can be preselected to eliminate some harmonics in the output voltage. This can be a very cost-effective means of reducing inverter-produced harmonics.

Since skin effect is responsible for the increased heating caused by harmonic currents, using conductors with larger surface areas will lessen the heating effects. This can be done by using finely stranded conductors, because the effective surface area of the conductor is the sum of the surface area of each strand.

Specially designed transformers called K-factor transformers are also advantageous when harmonic currents are prevalent. They parallel small conductors in their windings to reduce skin effect and incorporate special core designs to reduce the saturation effects at the higher flux frequencies produced by the harmonics.

Proper design needs to increase the size of neutral conductors to better accommodate triplen harmonics. Per the FPN in 210.4(A) and 220.22 of the 2002 NEC, "A 3-phase, 4-wire wye-connected power system used to supply power to nonlinear loads may necessitate that the power system design allow for the possibility of high harmonic neutral currents." And per 310.15(B)(4)(c), "On a 4-wire, 3-phase wye circuit where the major portion of the load consists of nonlinear loads, harmonic currents are present on the neutral conductor: the neutral shall therefore be considered a current-carrying conductor." It is important to note that the duct bank ampacity tables in B.310.5 through B.310.7 are designed for a maximum harmonic loading on the neutral conductor of 50% of the phase currents. Harmonics will undoubtedly continue to become more of a concern as more equipment that produces them is added to electrical systems. But if adequately considered during the initial design of the system, harmonics can be managed and their detrimental effects avoided.

It is well known that nonlinear loads like the switched-mode power supply (SMPS), variable frequency drives, electronic ballasts, and arc furnaces generate harmonic currents and voltages. Combining these with the nonlinear nature of the transformer core, waveform distortions in currents and voltages are created leading to an increase in power losses and winding temperature. In such cases, transformers supplying nonlinear devices should be derated based on the percentages of harmonic components in the rated winding eddy current loss and the load current. Another option is to use K-factor transformers. In other words, the need for looking into harmonic problems has become important.

4.5 Eddy Current Loss

Transformer losses are composed of copper and core losses, including the stray flux loss and eddy current. Eddy current loss is the power dissipated due to circulating currents in the core winding, as a result of electromotive forces induced by variation of magnetic flux, and it becomes considerable when harmonics exist. Harmonics tend to exponentially increase the transformer eddy current losses, causing higher operating temperature for the transformer. This occurs because eddy current loss is proportional to the square of the current in the conductor and the square of its frequency.

4.6 K-Factor

This is one way of establishing the capability of transformers to carry nonlinear loads. A K-factor of 1.0 means no harmonics. On the other hand, the presence of harmonic current gives a K-factor of more than 1.0. Basically, it is the sum of the product of the square of the harmonic currents and the square of the corresponding harmonic frequency number.

In equation form,

$$K = \left[\left(I_1/I\,\text{rms} \right)^2 (1)^2 \right] + \left[\left(I_2/I\,\text{rms} \right)^2 (2)^2 \right] + \left[\left(I_3/I\,\text{rms} \right)^2 (3)^2 \right] + \cdots$$

$$+ \left[\left(I_n/I\,\text{rms} \right)^2 (n)^2 \right] \tag{4.7}$$

where I_1 = fundamental current, I_2 = second harmonic current, I_3 = third harmonic current, I_n = nth harmonic current, and Irms = RMS current. Note that total RMS current is the square root of the sum of squares of the individual currents.

In addition, K-factor transformers are designed to tolerate K times the rated eddy current loss. In addition, these types of transformers have a larger neutral terminal, at least twice the size of the phase terminals, as protection against the triplen harmonics (3rd, 9th, 15th, etc.) that flows through the neutral.

4.7 Summary

To summarize, the effects of harmonic currents on transformers are

- Increased eddy current losses (ECLs)

- Additional copper losses due to the third-harmonic contents
- Electromagnetic interference with communication circuits

Meanwhile, harmonic voltages lead to the following:

- Increased dielectric stress on insulation (shortens insulation life)
- Resonance between winding reactance and feeder capacitance
- Electrostatic interference with communication circuits

Overall, the effect of harmonics is increased heating in the transformer as compared to purely sinusoidal operation. Furthermore, harmonics will result in lower efficiency, lesser capacity, reduced power factor, and decreased productivity.

4.8 Power Factor Control

Typically, a distributed PV system will maintain a lagging power factor greater than 0.9 when the output is greater than 50% of the rated inverter output power. This creates a significant amount of stress on the system considerations of DPV-GTs.

High-frequency noise is created by switching transients. There are two types of switching transients: the slow-switching transients and fast-switching transients. These are generated due to the electronic components in the inverters viz. the insulated gate bipolar transistors (IGBTs) and the gate drive circuit configurations. A typical diagram showing the components in a DPV-GT system is shown in Figure 4.1. The filters as shown in the circuit help to attenuate most of the harmonics and higher frequencies as experienced by the DPV-GT. DPV-GTs subjected to these harmonics have increased losses and thus can have lower efficiency. When the switching devices are turned on/off, the rapid and high dv/dt or di/dt changes cause very high oscillations of 100 kHz or higher. For higher current ratings, the integrated gate commutated/controlled thyristors (IGCTs) prove efficient and economical with a higher power handling capacity.

4.9 Shunt Filter

Generally, the resonant frequency of the shunt filter is around 150 kHZ which provides a harmonic and noise reduction in the bandwidth of 50 kHz to 5 MHz. This reduces the noise from the power stage as well as the switch mode stage in the DPV-GT system. This filter also protects the entire system from lightning impulses and current spikes of the order of kA in microseconds to voltages up to 1 kV.

4.10 Series Filter

The IGBT switches between 0.1 and 10 μsecs, which necessitates that the resonant frequency of the series filter be between 100 kHz and several MHz. Additionally, since the controller uses the switched-mode pulse scheme (SMPS) at 150 kHz, the series filter needs to be designed for attenuating both the common-mode (CM) and differential-mode (DM) noise. The attenuation for CM noise is at 80 dB for the frequencies between 100k HZ and 1 MHz and DM noise is at 70 dB between frequencies of 200 kHz and 3 MHz. The filter is also designed to eliminate the system dominant frequency components and is inactive in the lower range of the PWM frequency range.

4.11 Harmonic Mitigation

Power electronic devices that have rapid and frequent load variations have become abundant today due to their many process control related and energy saving benefits. However, they also bring a few major drawbacks to electrical distribution and DPV-GT systems, such as harmonics and rapid change of reactive power requirements.

Harmonics may disrupt normal operation of other devices and increase operating costs as illustrated in Figure 4.7. Symptoms of problematic harmonic levels include overheating of transformers, motors, and cables, thermal tripping of protective devices, and logic faults of digital devices and drives. Harmonics can cause vibrations and noise in electrical machines (motors, transformers, and reactors).

The life span of many devices can be reduced by elevated operating temperature.

There are various harmonic mitigation methods that we can use to address harmonics in the distribution system. They are all valid solutions depending on the circumstances, and they have both pros and cons. They are as follows: line reactors (LR)/DC bus chokes/isolation transformers, tuned harmonic filters, broadband filters, multipulse transformers/converters, and active harmonic filters (AHFs).

The concept of an active filter is to produce harmonic components, which cancel the harmonic components from the nonlinear loads. Figure 4.8

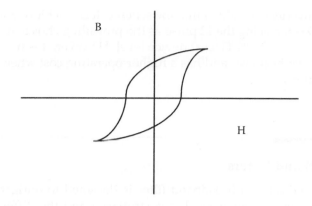

FIGURE 4.7
B (Flux density in tesla) versus H (magnetic field intensity in AT/m)—magnetization curve of core material.

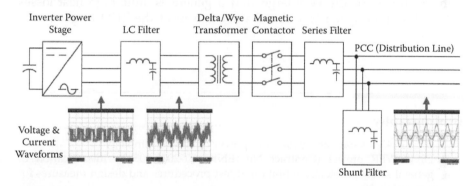

FIGURE 4.8
(See color insert) Circuit for harmonic mitigation.

illustrates how the harmonic current generated by AHF is injecting into the system to cancel the harmonic from a variable frequency drive (VFD) load. An AHF is a highly effective device that cancels multiple order harmonics in the distribution system. It is installed as a parallel device and scaled via paralleling multiple units. It can handle different types of loads, linear or nonlinear. It addresses harmonics from a system point of view and can save significant cost/space in many applications. Its performance level can meet total demand distortion (TDD) 5% target.

The 12 or 18 pulses variable frequency drive (VFD) has been developed in industry to address the harmonic issue caused by common 6 pulse VFD. Figure 4.8 presents a typical concept of 12 pulse VFD. The input is connected to the transformer's primary winding, and then the outputs are connected with two separated phase-shifted secondary windings to two sets of inverters. This configuration reduces the current harmonic distortion to a 10% range (12 pulse). For the 18 pulse VFD, an additional secondary winding and a set of inverters are added into the scheme. It can achieve 5% TDD. The 18 pulse VFD is replacing the 12 pulse as the prevailing choice in multipulse solutions. It can reach 5% TDD at device level. However, it is normally very bulky, has larger heat loss, and has a higher operating cost when compared to other solutions.

4.12 Broadband Filters

As its name indicated, a broadband filter is designed to mitigate multiple orders of harmonic frequencies. It has similarity and the difference of its circuit from the tune filter. Both inductors (L) could have an impedance >8%, which means there will be a 16% voltage drop across the filter. Its physical dimension is normally very large, and it generates quite high heat losses (>4%). A well-designed broadband filter can meet the TDD target in the 10% range.

Bibliography

1. IEEE P1433, A standard glossary of power quality terminology.
2. DISPOWER project (Contract No. ENK5-CT-2001-00522), Identification of general safety problems, definition of test procedures and design-measures for protection, 2004.
3. DISPOWER project (Contract No. ENK5-CT-2001-00522), Summary report on impact of power generators distributed in low voltage grid segments, 2005 (http://www.pvupscale.org).
4. IEA PVPS Task V, report IEA-PVPS T5-01: 1998, Utility aspects of grid connected photovoltaic power systems.
5. IEC 61000-3-2: 2005, EMC—Part 3-2: Limits—Limits for harmonic current emissions equipment input current up to and including 16 A per phase.
6. IEEE 929: 2000, Recommended practice for utility interface of photovoltaic (PV) systems.
7. Engineering Recommendation G77/1: 2000, Connection of single-phase inverter connected photovoltaic (PV) generating equipment of up to 5 kW in parallel with a distribution network operator (DNO) distribution system.
8. IEC/TS 61000-3-4:, EMC—Part 3-4: Limits—Limitation of emission of harmonic currents in low-voltage power supply systems for equipment with rated current greater than 16 A.
9. IEC/TR3 61000-3-6:, EMC—Part 3-6: Limits—Assessment of emission limits for distorting loads in MV and HV power systems—Basic EMC publication.
10. Cobben, J. F., Heskes, P. J., Moor de, H. H., Harmonic distortion in residential areas due to large scale PV implementation is predictable. *DER-Journal*, January 2005.

11. Cobben, J. F., Kling, W. L., Heskes, P. J., Oldenkamp, H., Predict the level of harmonic distortion due to dispersed generation, 18th International Conference on Electricity Distribution (CIRED), Turin, Italy, June 2005.

12. Cobben, J. F., Kling, W. L., Myrzik, J. M., Making and purpose of harmonic fingerprints, 19th International Conference on Electricity Distribution (CIRED), Vienna, Austria, May 2007.

13. Oldenkamp, H., De Jong, I., Heskes, P. J. M., Rooij, P. M., De Moor, H. H. C., Additional requirements for PV inverters necessary to maintain utility grid quality in case of high penetration of PV generators, 19th EC PVSEC (2004), pp. 3133–3136.

14. Cobben, J. F. G., Power quality implications at the point of connection, Dissertation, University of Technology Eindhoven, 2007.

15. IEA-PVPS Task V, report IEA-PVPS T5-2: 1999, Demonstration test results for grid interconnected photovoltaic power systems.

16. Halcrow Gilbert Associates, Department of Trade and Industry, Coordinated experimental research into power interaction with the supply network—Phase 1 (ETSU S/P2/00233/REP), 1999 (http://www.dti.gov.uk/publications).

17. UNIVERSOL project (Contract No. NNE5-293-2001), Quality impact of the photovoltaic generator "Association Soleil-Marguerite" on the public distribution network, EDF-R&D, 2004.

18. IEC 61000-4-7: 2002, Electromagnetic compatibility (EMC)—Part 4-7: Testing and measurement techniques—General guide on harmonics and interharmonics measurements and instrumentation, for power supply systems and equipment connected thereto.

19. IEC 61000-2-12: 2003. Electromagnetic compatibility (EMC)—Part 2-12: Environment—Compatibility levels for low-frequency conducted disturbances and signaling in public medium voltage power supply systems.

20. Hong, Soonwook, Zuercher–Martinson, Michael, Harmonics and noise in photovoltaic (PV) inverter and the mitigation strategies, white paper, Solectria, Lawrence, MA.

21. Harmonic analysis report, Multiple loads, Allied Industrial Marketing, September 2011, Cedarsburg, WI.

Chapter 4 Problems

1. Design an LC filter for the PWM harmonic mitigation system in Figure 4.6 with an LC filter resonant frequency of 750 Hz. The PWM modulation frequency is 10 kHZ to be attenuated at 45 dB below the fundamental component. Conduct the FFT analysis and draw the Bode plot to illustrate your design of the LC filter. Show that the 10-kHZ ripple component is further attenuated to 60 dB below the

fundamental component of the shunt filter. What is the percentage of this ripple to the fundamental component and the overall TDD? Calculate the THD and TDD for this LC filter.

2. Design a resistance inductor capacitor (RLC) shunt filter whose resonant frequency is 150 kHz with a BW of 50 kHz to 5 MHz. The response to a current pulse of $I = 0A$, $I = \pm 1$ A, and $I = \pm 10$ A should be within 2 to 5 µs. Calculate the THD and TDD for this RLC shunt filter.

3. Design a RLC series filter so that the filter is tuned between 100 kHZ and 5 MHz. The SMPS controller is switched at a frequency of 150 kHz. In addition, the series filter should attenuate the CM noise at 80 dB for the frequencies between 100k HZ and 1 MHz and DM noise at 70 dB between frequencies of 200 kHz and 3 MHz. Calculate the THD and TDD for this RLC series filter.

5

Frequency Variation, Power Factor Variation in Distributed Photovoltaic Grid Transformers

5.1 Under- or Over-Frequency

Frequency variations in the grid require a response from the PV system to ensure the safety of equipment connected to the point of coupling, whether it is a photovoltaic (PV) system, utility network, or consumer equipment. The PV system must operate synchronously with the utility system within ±1 Hz as per international standards and cease to energize the utility line within 0.2 seconds when outside the above range. The utility system must again be synchronized with the PV system when frequency settles within 60 Hz ±1. Interconnecting to existing under-frequency load-shed (UFLS) blocks is generally avoided. Such an action prevents tripping significant generation along with load.

5.2 Power Factor Control

Typically, a distributed PV system will maintain a lagging power factor greater than 0.9 when the output is greater than 50% of the rated inverter output power. This creates a significant amount of stress on the system considerations of distributed photovoltaic grid transformers (DPV-GTs).

5.3 Under-Frequency Concerns

Generally, interconnecting to existing under-frequency load-shed (UFLS) blocks is avoided; this prevents tripping significant generation along with load. Islanding also causes changes in frequency. Depending on the acceptable or permissible value changes in the real and reactive parts of the apparent power, the frequency changes will move out of the permissible limits. With voltage and frequency variations matched by balancing the generation and load demands leading to a stable islanded system, typically such a system is characterized by what is known as the nondetection zone (NDZ). This aspect is described in detail in Chapter 6 and is briefly illustrated here.

5.4 Over-/Under-Voltage (OVP/UVP) and Over-/Under-Frequency (OFP/UFP)

Grid-tied DPV-GT systems need to have OVP/UVP and/or OFP/UFP detection systems that will cause the PV inverter to stop supplying power to the DPV-GT and the grid if the frequency or the amplitude of the voltage at the point of common coupling (PCC) between the customer (load) and the utility vary beyond prescribed limits. These serve to protect the customer's equipment and also serve as anti-islanding detection methods.

In Figure 5.1 let the real power flowing into the PCC from the utility be

$$\Delta P = P_{\text{load}} - P_{\text{g}} \tag{5.1}$$

And the reactive power flowing into PCC from the utility be

$$\Delta Q = Q_{\text{load}} - Q_{\text{g}} \tag{5.2}$$

The corresponding apparent power components from the PV generator into the PCC and apparent power from PCC to load are shown in Figure 5.2. If the PV inverter power operates at upf, $Q_{\text{g}} = 0$ and $\Delta Q = Q_{\text{load}}$. Further, the behavior

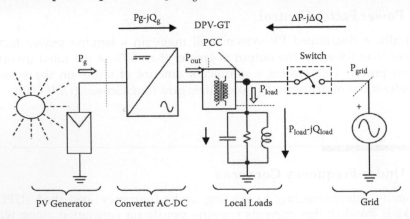

FIGURE 5.1
A typical representation of islanding in a DPV-GT system with OVP/UVP or OFP/UFP method.

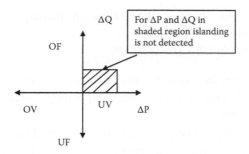

FIGURE 5.2
Typical mapping of NDZ in ΔP versus ΔQ region for an OVP/UVP and OFP/UFP islanding detection scheme.

of the entire DPV-GT system will depend on ΔP and ΔQ a few instants before the switch opens to form the island. Further, if $\Delta P \neq 0$, the amplitude of voltage at PCC will change, and the OVP/UVP can detect the change and prevent islanding. For any such islanding activity that is not averted, the DPV-GT will be subjected to changes in voltage levels which will require proper attention for insulation coordination. If $\Delta Q \neq 0$, the load voltage will show a sudden shift in phase, which will cause the inverter control system to cause a shift in the frequency of the inverter current and in turn the frequency of the voltage at load, until $\Delta Q = 0$ (i.e., the load resonant frequency is reached). This change in frequency can be detected by the OFP/UFP. This necessitates the requirement of an efficient phase-locked-loop (PLL) system to maintain appropriate resonant frequency close to the one specified by the utility. A slower PLL circuit will cause a step-phase shift in the voltage equal to the power factor. Many stringent OVP/UVP and OFP/UFP metrics are generally in use or are specified by an utility. Therefore, if either the real power of the load and PV system (inverter output) is not matched, *or* the load's resonant frequency does not lie near the utility frequency, islanding will not occur. The strength of such an anti-islanding system is the fact that OVP/UVP and OFP/UFP are used for several other reasons and ultimately the system is effective for the deactivation of the inverter circuit and is a low-cost option for detecting islanding. It also helps to protect the DPV-GT. At the same time the weakness of such a system in terms of prevention of islanding is the large NDZ. A typical NDZ metric is illustrated in Figure 5.2.

5.5 Frequency Variation due to Electromagnetic Compatibility (EMC)

Both the International Electric Commission (IEC) and the European Union (EU) define EMC to cover electromagnetic phenomena from zero hertz.

Furthermore, the IEC defines the following principal electromagnetic-conducted phenomena:
Conducted low-frequency phenomena:

- Harmonics, interharmonics—the odd and even harmonics can contribute to frequency variations, especially those caused by the second harmonic.
- Signals superimposed on power lines—the third harmonic creates a nuisance factor in the signal carried especially by the communication circuits that generally have the same right of way as the distribution system.
- Voltage fluctuations
- Voltage dips and interruptions
- Voltage unbalance
- Power frequency variations
- Induced low-frequency voltages
- DC component in AC networks

5.6 Conducted High-Frequency Phenomena

- Induced voltages or currents—these further increase the ECL and hysteresis losses
- Unidirectional transients
- Oscillatory transients—make insulation coordination a difficult proposition and can hamper efficient design of the DPV-GT. Clever and intelligent design procedures and insulation design procedures need to be followed with proper shunts to reduce the stray losses as well.

5.7 Frequency Problems Related to Large Grid Tied DPV-GT Impedance and PV Inverter Interaction

- Photovoltaic (PV) inverters used in dispersed power generation of residential DPV systems are presently, generally in the range of 1 to 5 kW and are available from several manufacturers. However, large grid impedance variation is challenging the control and the grid filter design in terms of stability. In fact the PV systems are well suited for

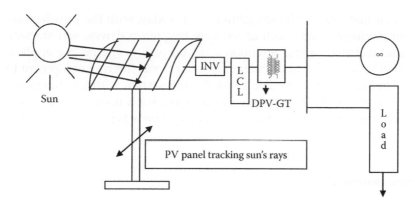

FIGURE 5.3
LCL filter with proper resonant frequency.

loads connected in a great distance to the transformer (long wires), and the situation becomes even more difficult in low-developed remote areas characterized by low power transformers and long distribution wires with high grid impedance. Hence, a theoretical analysis is needed because the grid impedance variation leads to dynamic and stability problems in both the low-frequency range (around the current controller bandwidth frequency) as well as the high-frequency range (around the LCL-filter resonance frequency). A typical LCL is shown in Figure 5.3. The choice of the resonant frequency is given by solution of the T-shaped LCL filter.

In the low-frequency range, the possible variation of the impedance challenges the design of resonant controllers adopted to mitigate the effect of the grid harmonic distortion on the grid current. In the high-frequency range, the grid impedance influences the frequency characteristic of the filter, and the design of passive or active damping (to ensure stability) becomes more difficult. Both topics are addressed and discussed here with simulation and experimental results.

5.8 Power Factor Correction (PFC)

- The power factor (PF) can be improved by either increasing the active power component or reducing the reactive component. For a given load, increasing the active power component for the sole purpose of power factor correction would not be economically feasible. Thus, the only practical means for improving the system's power factor is to reduce the reactive power component. Applying power factor capacitors used to be straightforward. Today, with the proliferation of nonlinear loads such as variable frequency drives, soft starters, and welders, careful attention must be paid to the proper application of power factor correction and harmonic filtering equipment to avoid misapplication. For a distribution system with low harmonic content, standard capacitors can be used. For a high harmonic content environment, a detune capacitor system is typically required.

Bibliography

1. IEEE P1433, A standard glossary of power quality terminology.
2. DISPOWER project (Contract No. ENK5-CT-2001-00522), Identification of general safety problems, definition of test procedures and design-measures for protection, 2004.
3. DISPOWER project (Contract No. ENK5-CT-2001-00522), Summary report on impact of power generators distributed in low voltage grid segments, 2005 (http://www.dispower.org).
4. IEA PVPS Task V, report IEA-PVPS T5-01: 1998, Utility aspects of grid connected photovoltaic power systems.
5. IEC 61000-3-2: 2005, EMC—Part 3-2: Limits—Limits for harmonic current emissions equipment input current up to and including 16 A per phase.
6. IEEE 929: 2000, Recommended practice for utility interface of photovoltaic (PV) systems.
7. Engineering Recommendation G77/1: 2000, Connection of single-phase inverter connected photovoltaic (PV) generating equipment of up to 5 kW in parallel with a distribution network operator (DNO) distribution system.
8. IEC/TS 61000-3-4:,EMC—Part 3-4: Limits—Limitation of emission of harmonic currents in low-voltage power supply systems for equipment with rated current greater than 16 A.
9. IEC/TR3 61000-3-6: EMC—Part 3-6: Limits—Assessment of emission limits for distorting loads in MV and HV power systems—Basic EMC publication.
10. Cobben, J. F., Heskes, P. J., Moor de, H. H., Harmonic distortion in residential areas due to large scale PV implementation is predictable, *DER-Journal*, January 2005.
11. Cobben, J. F., Kling, W. L., Heskes, P. J., Oldenkamp, H., Predict the level of harmonic distortion due to dispersed generation, 18th International Conference on Electricity Distribution (CIRED), Turin, Italy, June 2005.
12. Cobben, J. F., Kling, W. L., Myrzik, J. M., Making and purpose of harmonic fingerprints, 19th International Conference on Electricity Distribution (CIRED), Vienna, Austria, May 2007.
13. Oldenkamp, H., De Jong, I., Heskes, P. J. M., Rooij, P. M., De Moor, H. H. C., Additional requirements for PV inverters necessary to maintain utility grid quality in case of high penetration of PV generators, 19th EC PVSEC (2004), pp. 3133–3136.

14. Cobben, J. F. G., Power quality implications at the point of connection, Dissertation, University of Technology Eindhoven, 2007.
15. IEA-PVPS Task V, report IEA-PVPS T5-2: 1999, Demonstration test results for grid interconnected photovoltaic power systems.
16. Halcrow Gilbert Associates, Department of Trade and Industry, Coordinated experimental research into power interaction with the supply network—Phase 1 (ETSU S/P2/00233/REP), 1999 (http://www.dti.gov.uk/publications).
17. UNIVERSOL project (Contract No. NNE5-293-2001), Quality impact of the photovoltaic generator "Association Soleil-Marguerite" on the public distribution network, EDF-R&D, 2004.
18. IEC 61000-4-7: 2002, Electromagnetic compatibility (EMC)—Part 4-7: Testing and measurement techniques—General guide on harmonics and interharmonics measurements and instrumentation, for power supply systems and equipment connected thereto.
19. IEC 61000-2-12: 2003. Electromagnetic compatibility (EMC)—Part 2-12: Environment—Compatibility levels for low-frequency conducted disturbances and signaling in public medium voltage power supply systems.
20. Hong, Soonwook, Zuercher–Martinson, Michael, Harmonics and noise in photovoltaic (PV) inverter and the mitigation strategies, white paper, Solectria, Lawrence, MA.
21. IEA PVPS Task V, report IEA-PVPS T5-11: 2002, Grid connected photovoltaic power systems: Power value and capacity value of PV systems.
22. Perez, R., et al., Photovoltaics can add capacity to the utility grid, Report NREL-DOE/GO-10096-262, 1998.
23. Perez, R., Schlemmer, J., Bailey, B., Elsholz, K., The solar load controller end-use maximization of PV's peak shaving capability, *Proceedings of the American Solar Energy Society Conference*, 2000.
24. EA Technology, Department of Trade and Industry, Overcoming barriers to scheduling embedded generation to support distribution networks (ETSU K/EL/00217/REP), 2000.
25. Perez, R., Letendre, S., Herig, C., PV and grid reliability: Availability of PV power during capacity shortfalls, *Proceedings of the American Solar Energy Society Conference*, 2001.
26. Perez, R., et al., Availability of dispersed photovoltaic resource during the August 14th, 2003, northeast power outage, *Proceedings of the American Solar Energy Society Conference*, 2004.

6

Islanding Effects on Distributed Photovoltaic Grid Transformers

Islanding refers to the condition of a distributed generation (DG) generator continuing to power a location even though power from the electric utility is no longer present. Consider for example a distributed solar facility that has solar panels that feed power back to the electrical grid; in the case of a power blackout, if the solar panels continue to power the distributed solar facility, the solar facility becomes an "island" with power surrounded by a "sea" of unpowered distributed solar facilities in a distributed solar grid system. This condition may prove dangerous and sometimes fatal to humans if not properly monitored and controlled.

There are two types of islanding:

1. *Operational islanding* is also called a self-supporting power system capable of delivering power from supplier to customer within acceptable limits of voltage and frequency variation. Such islands are parts of normal operation of such standard networks. Present grid operations necessitate the provision of such islands.

2. In *unintentional islanding,* the generation is disconnected from the network automatically within a certain disconnect time to protect network operators and distributed photovoltaic grid transformers (DPV-GTs) and other equipment from damage. This is as described in the introductory paragraph and is the main topic of discussion.

Islanding can be dangerous to utility workers, who may not realize that the solar facility is still powered even though there is no power from the grid. For that reason, distributed generators must detect islanding and immediately stop producing power. Mechanisms to interject and interrupt such events from occurring have now been developed by the industry.

In *intentional islanding,* the customer disconnects the building from the grid and forces the distributed generator to power the building or the load. Methods to detect islanding are described below with the nondetection zone (NDZ) as one of the metrics to assess the weakness or use of any method. NDZ helps to assess the weakness of an islanding detection method to fail to detect an islanding scenario.

A typical PV system with DPV-GT is shown in Figure 6.1, where a PV generator is connected to an inverter to a point of common coupling (PCC)

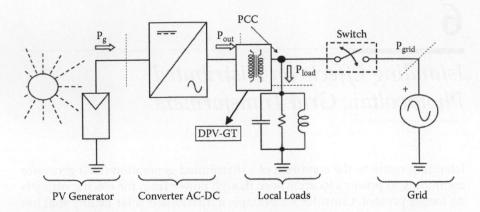

FIGURE 6.1
Islanding in a DPV-GT system.

through a DPV-GT, followed by local loads and a switch that ties it to the grid. Typically RLC loads with a high value of Q factor pose severe problems for islanding detection. The Q factor is defined as

$$Q = R\left\{SQRT\left(C/L\right)\right\} \tag{6.1}$$

Most of the islanding prevention methods generally involve high Q RLC loads. The introduction of DPV-GT in the circuit adds to the value of inductance besides providing an additional resistance of the inductive transformer circuit. High Q loads generally have large capacitances and small values of inductances and related resistances. At the same time, harmonic producing loads or constant power resistive loads do not pose such problems in islanding detection.

6.1 EN61000-3-2 European Standard Regulating Harmonic Currents

The corresponding IEC standard 61000-3-2 imposes limits on the harmonic currents drawn from the main supply. This standard requires that electrical appliances be type tested to ensure that they meet the requirements in the standard.

This is applicable to electrical and electronic equipment having an input current up to and including 16A per phase and intended to be connected to public low-voltage distribution systems (i.e., supply voltages nominally 230V AC or 415V AC three phase).

The standard defines four classes of waveform according to the different types of equipment. For example, one of the Class B applies to portable tools, whereas the typical switched-mode waveform is generally in Class D. Each class has different harmonic limits up to the 40th, which must not be exceeded. Some classes have dynamic limits that are set according to the power drawn by the device.

The scope of the EN61000-3-2 standard includes products such as lighting equipment, portable tools, all electronic equipment, consumer products and appliances, and industrial equipment. This standard does not cover equipment that has a nominal supply voltage less than 220V AC. No limits have been specified for professional equipment above 1 kW.

Although these requirements cover only products to be sold within EEC countries, a similar IEEE document exists for the United States, and Japan is also considering similar legislation. UL 1741 provides some safety guidelines.

This international standard applies to utility-interconnected PV power systems operating in parallel with the utility and utilizing static (solid-state) non-islanding inverters for the conversion of DC to AC. This document describes specific recommendations for systems rated at 10 kVA or less, such as may be utilized on individual residences as single or three phase. This standard applies to interconnection with the low-voltage utility distribution system.

The object of this standard is to lay down requirements for the interconnection of PV systems to the utility distribution system.

NOTE 1: An inverter with type certification meeting the standards as detailed in this standard should be deemed acceptable for installation without any further testing.

This standard does not deal with electromagnetic compatibility (EMC) or protection mechanisms against islanding.

NOTE 2: Interface requirements may vary when storage systems are incorporated or when control signals for PV system operation are supplied by the utility.

6.2 Scoping Consistency

Generally a utility will develop a scoping template to maintain a study consistency. The aspects approached in such studies are as follows:

1. A threshold methodology for 15% peak load is presented. Peak load is monitored and recorded by a utility for all distribution circuits. The minimum load has not traditionally been monitored, yet has been made available in recent years on a few installations via remote metering. The minimum load on radial circuits is typically 30% of the peak load. The data sampling period should be at least one year and should represent typical system loading conditions.

2. Islanding becomes a power quality and protection concern when the maximum aggregate generation on a radial distribution circuit approaches 50% of the minimum load. Therefore, planning engineers use 15% of peak load screen (50% of 30%) to quickly identify if the interconnection request poses a potential islanding and power quality concern. If the interconnection request and aggregated generation exceed 15% of peak load, the utility can re-evaluate this screen using minimum load if available as opposed to peak load. Failing this screen a second time would indicate that further study is necessary and special protection requirements may result to allow the interconnection.

6.3 Methods for Detecting Islanding with DPV-GTs That Are Grid-Tied

Islanding can be detected by passive and active methods resident in the inverter, and by methods not resident in the inverter but present at the utility levels.

Passive methods are described and evaluated below [1–3].

6.3.1 Over-/Under-Voltage (OVP/UVP) and Over-/Under-Frequency (OFP/UFP)

Grid-tied DPV-GT systems need to have OVP/UVP and/or OFP/UFP detection systems that will cause the PV inverter to stop supplying power to the DPV-GT and the grid if the frequency or the amplitude of the voltage at the PCC between the customer (load) and the utility vary beyond prescribed limits. These serve to protect customer's equipment and also serve as anti-islanding detection methods.

In Figure 6.2 let the real power flowing into the PCC from the utility be

$$\Delta P = P_{\text{load}} - P_{\text{g}} \tag{6.2}$$

And let the reactive power flowing into PCC from the utility be

$$\Delta Q = Q_{\text{load}} - Q_{\text{g}} \tag{6.3}$$

The corresponding apparent power components from PV generator into the PCC and apparent power from PCC to load are shown in Figure 6.2. If the PV inverter power operates at unity power factor (upf), $Q_{\text{g}} = 0$ and $\Delta Q = Q_{\text{load}}$. Further, the behavior of the entire DPV-GT system will depend on ΔP and ΔQ a few instants before the switch opens to form the island. Further, if $\Delta P \neq 0$, the amplitude of voltage at PCC will change and the OVP/UVP can detect the

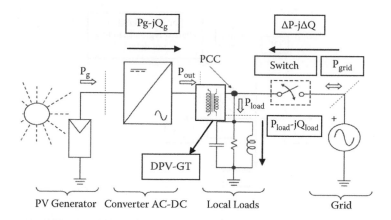

FIGURE 6.2
Islanding in a DPV-GT system with OVP/UVP or OFP/UFP method.

change and prevent islanding. Any such islanding activity that is not averted will subject the DPV-GT to changes in voltage levels that will need proper attention for insulation coordination. If $\Delta Q \neq 0$, the load voltage will show a sudden shift in phase, which will cause the inverter control system to cause a shift in the frequency of the inverter current and in turn the frequency of the voltage at load, until $\Delta Q = 0$ (i.e., the load resonant frequency is reached). This change in frequency can be detected by the OFP/UFP. This necessitates the requirement of an efficient phase-locked-loop (PLL) system to maintain appropriate resonant frequency close to that specified by the utility. A slower PLL circuit will cause a step-phase shift in the voltage equal to the power factor. Many stringent OVP/UVP and OFP/UFP metrics are generally in use or specified by a utility. Therefore, if either the real power of the load and PV system (inverter output) is not matched, *or* the load's resonant frequency does not lie near the utility frequency, islanding will not occur. The strength of such an anti-islanding system is the fact that OVP/UVP and OFP/UFP are used for several other reasons and ultimately are effective for the deactivation of the inverter circuit and are a low-cost option for detecting islanding. The system also helps to protect the DPV-GT. At the same time the weakness of such a system in terms of the prevention of islanding is the large NDZ. A typical NDZ metric is illustrated in Figure 6.3.

6.3.2 Voltage-Phase Jump Detection (PJD)

This is similar to power factor detection or transient-phase detection. The difference in phase between the inverter current and the voltage at PCC is monitored for a sudden jump. Under normal operation and with current-source inverters, the inverter current is synchronized with the voltage at the PCC by using the rising or falling zero crossings aided by a suitable PLL device. Once the islanding takes place, the voltage jumps and the phase shift are detected.

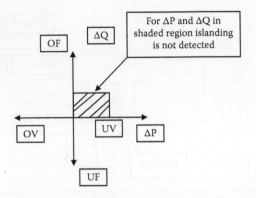

FIGURE 6.3
Typical mapping of NDZ in ΔP versus ΔQ region for an OVP/UVP and OFP/UFP Islanding detection scheme.

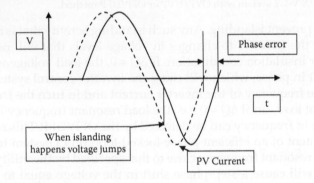

FIGURE 6.4
Phase-shift jump related to the PJD scheme; phase error is caused by voltage jump at islanding.

This causes a phase difference between the voltage and the inverter current. If this error is greater than a preset threshold, islanding can be avoided. PJD is simple to implement but the threshold selection is tricky. Sensitivity of the NDZ to cause nuisance trips cannot be easily controlled, especially when the load has a phase angle of zero, as it will not create a phase error when the utility is disconnected, as shown in Figure 6.4.

6.3.3 Detection of Voltage Harmonics

This is similar to the detection of impedance at a particular frequency. The PV inverter monitors the third harmonic distortion (THD), and if this exceeds a certain threshold, the PV inverter is deactivated. When an island occurs, two mechanisms can cause the voltage at the PCC to introduce harmonics:

1. PV inverter
2. DPV-GT

When an island occurs, there are two mechanisms that can cause the harmonics in voltage at PCC to increase. One of these is the PV inverter itself. A PV inverter will produce some current harmonics in its AC output current, as all switching power converters do. A typical requirement for a grid-connected PV inverter is that it produces no more than 5% THD of its full rated current [11,12]. When the utility disconnects, the harmonic currents produced by the inverter will flow into the load, which in general has much higher impedance than the utility. The harmonic currents interacting with the larger load impedance will produce larger harmonics in voltage at PCC [8]. These voltage harmonics, or the change in the level of voltage harmonics, can be detected by the inverter, which can then assume that the PV inverter is islanding and discontinue operation.

The second mechanism that may cause the harmonics to increase is the voltage response of the transformer shown in Figure 6.1. When current-source inverters are used, and when the switch that disconnects the utility voltage source from the island is on the primary side of the transformer, as shown in Figure 6.2, the secondary of the transformer will be excited by the output current of the PV inverter. However, because of the magnetic hysteresis and other nonlinearities of the transformer, its voltage response is highly distorted [8] and will increase the THD in voltage at PCC. There can also be nonlinearities in the local load, such as rectifiers, which would similarly produce distortion in voltage at PCC. These nonlinearities tend to produce significant third harmonics in general. Thus, when this method is used in practice, it is frequently the third harmonic that is monitored. This method fails when the load has low-pass characteristics. This method also is prone to fail when loads at the generator shunt reactive currents. This method may also fail when inverters have high-quality, low-distortion outputs.

6.3.4 Detection of Current Harmonics

The PV inverter will also create some current harmonics that can be monitored using a similar technique.

The active detection methods for islanding that are typically resident in the inverter generally contain an active circuit to force voltage, frequency, or the measurement of impedance, as discussed in the following sections.

6.3.4.1 Impedance Measurement

These are also called power shift, current notching, or output variation. In Figures 6.1 and 6.2 the PV inverter can appear as a current source given by

$$i_{\text{PV-inv}} = I_{\text{PV-inv}} \sin\left(\omega_{\text{PV}} t + \Phi_{\text{PV}}\right) \tag{6.4}$$

There are three parameters that can be varied: $I_{\text{PV-inv}}$, ω_{PV}, and Φ_{PV}. In the output variation the variation is imposed on the amplitude of the PV-Inv current. When connected to the utility through the DPV-GT, the voltage perturbation resulting from the current amplitude variation, ultimately also resulting in the power perturbation, is given by

$$\Delta V = (\Delta P/2) \cdot \text{Sqrt}(R/P) \tag{6.5}$$

When the grid is disconnected it will result in a variation in the voltage at the PCC that can be used to prevent islanding. If the analysis is evaluated, this variation is basically a change in the impedance as shown below:

$$Z = dV_{PCC}/dI_{PV\text{-}inv} \tag{6.6}$$

Equation (6.6) is basically the impedance evaluated by the inverter, and thus is given the name *impedance method* [1]. Variations in V_{PCC} can be achieved by loading the inverter to the limit of the UVP/OVP. The minimum current shift needed to detect islanding is equal to the full size of the UVP/OVP window. Thus, for a 10% V_{PCC} change, a 20% change in $I_{PV\text{-}inv}$ is necessary. This method is highly recommended for a single inverter configuration, because the NDZ is comparatively small with any local load where the load impedance is larger than the grid impedance. When the gird disconnects and the power of the inverter and the load are balanced, the variation in the inverter voltage will cause the UVP to trip. However, with multiple inverters the effectiveness of the impedance method fails because the perturbations in the resultant voltage at PCC and its variation are not enough to cause the UVP to trip.

Example 6.1

In Figure 6.5a, a single inverter's output is shown. In Figure 6.5b, the output of 50 nonsynchronized inverters is shown.

This system is designed to reduce its output power by 20% every 20 time-units. This single inverter probably would not island because the 20% power drop would most likely lead to a large enough drop in voltage to trip the UVP. However, Figure 6.5b shows the power production of 50 inverters, all identical to the one in Figure 6.5a except that the 20% power "perturbations" are not synchronized. The maximum variation from the mean power production of the 50 PV inverters is less than 2%, and the UVP will probably no longer detect a trip condition. For a case with higher impedance, the NDZ window will have to be set higher.

6.3.4.2 Detection of Impedance at a Specific Frequency

This is also called the harmonic amplitude jump method and is a special case of the harmonic detection method. It is active because a current harmonic of a particular frequency is injected into the PCC by the inverter. When the grid is connected and if the grid impedance is smaller than the load impedance, the harmonic current at a particular frequency flows through

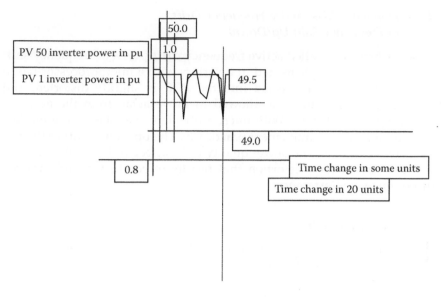

FIGURE 6.5
Failure of 50 inverter nonsynchronized with impedance method.

the grid and no voltage variation is noticed. When the grid disconnects, this current flows through the load. If the load is linear (i.e., like a parallel combination of RLC), then it is possible to inject a harmonic current into the PCC. Then a voltage particular to this harmonic can be detected to initiate the trip. Since the harmonic voltage is proportional to the impedance of the load and the harmonic current at a particular frequency, it is called the impedance detection at a specific frequency. This method has similar characteristics as the harmonic detection method. The NDZ space is similar to the harmonic detection method as well. A subharmonic current can be injected, but the grid/utility may have problems with this injection.

6.3.4.3 Slip-Mode Frequency Shift (SMS)

This method is also called the slide-mode frequency shift, PLL slip, or the follow-the-herd method. SMS is used to apply positive feedback to destabilize the inverter when the grid is not connected, thus preventing a steady-state long run-on. As shown in Equation (6.4), all three parameters of the voltage at PCC can be subjected to positive feedback: amplitude, frequency, and phase. SMS is used to apply positive feedback to the phase of V_{PCC} so as to shift the phase and thus the short-term frequency. The frequency of the grid is not affected by this method. This method is easy to implement and has a small NDZ, besides being applicable to multiple inverters in the DPV-GT system. But, this method can reduce the power quality and introduce transient response problems.

6.3.4.4 Frequency Bias (Active Frequency Drift or Frequency Shift Up/Down)

Frequency bias (also called active frequency drift [AFD] or frequency shift up/down) is easily implemented in inverters with a microprocessor-based controller. The current injected into the PCC is slightly distorted, and the frequency drift caused is allowed to move away from the ω_0 to be detected by the OFP/UFP. With microcontrollers available in the market, this method is easy to implement practically. The up/down shift of the frequency bias can cause the power quality to be affected drastically, causing discontinuous current distortion that results in conducted and radiated radio-frequency interference.

6.3.4.5 Frequency Shift

The traditional frequency-shift methods for islanding detection of grid-connected DPV-GTs and PV inverters are

1. The active frequency drift method
2. The slip-mode frequency-shift method

The above two can become ineffective under certain paralleled RLC loads. The automatic phase-shift method is used to alleviate this problem. This method is based on the phase shift of the sinusoidal inverter output current. When the utility malfunctions, the phase-shift algorithm keeps the frequency of the inverter terminal voltage deviating until the protection circuit is triggered.

6.3.4.6 Voltage Shift (VS)

This is also called the positive on the voltage or the follow-the-herd method. As suggested, this is the third method where a positive feedback is applied to the voltage at PCC. If there is a decrease in the root-mean-square (rms) value of V_{PCC}, the inverter reduces the amplitude of the voltage and thus the reduction in the power supplied by the inverter as well. However, if the utility is connected, this makes little difference when the power is reduced. If the utility is disconnected and there is a reduction in the amplitude of V_{PCC}, then by Ohm's law response by the RLC circuit, this will cause a reduction in the current which in turn causes a reduction in the PV inverter current with an eventual reduction in the amplitude of V_{PCC} that can be detected by the UVP scheme. There is a possibility of increasing or decreasing the PV inverter power which can be detected by the OVP/UVP scheme, but the trip created by the UVP is much more appealing because the disconnect will prevent the damage caused to the load equipment. This scheme works for single or multiple inverter schemes with OVP/UVP or OFP/UFP mechanisms in operation. This VS scheme is highly effective with the UVP and UFP in tandem for multiple inverters with microcontroller-based methods in the inverter being easy to implement. The VS and FS methods are highly effective in preventing islanding in DPV-GT systems with an NDZ. The NDZ for this scheme is similar to the one for the four-quadrant scheme of OVP/UVP and OFP/UFP, but the overall NDZ area is comparatively smaller and the Q of the load has insignificant effect on the operation of this method.

6.3.4.7 Frequency Jump (FJ)

This is also called the zebra method and is similar to the frequency bias method. The methodology is similar to the impedance measurement method. Dead zones are inserted in the output current in this FJ method but not in every cycle. Thus the frequency of output current is occasionally dithered according to a specified pattern. This could happen in the fifth cycle and in some cases a sophisticated pattern scheme is implemented. If connected to the utility, the frequency jump causes the inverter current to vary but with domination from the utility grid voltage at PCC. When disconnected from the grid, the FJ method causes a detection of islanding in the same manner as the frequency bias method or by detection of the frequency of the voltage at PCC that matches the dithering pattern used by the inverter. If the FJ method pattern is sophisticated, it can serve as a very strong anti-islanding scheme for single-inverter DPV-GT systems and fails if multiple inverters are used because each individual inverter scheme can cancel each other out, causing failure of detection by the FJ method. Thus, this scheme has absolutely zero NDZ for a single-inverter case but loses effectiveness in a multiple-inverter scheme.

6.3.4.8 ENS or MSD (Device Using Multiple Methods)

Very intelligent ways of integrating into a single scheme can be used by integrating an adaptive system similar to an adaptive filtering scheme or an observer method. A considerable amount of research is still being done to implement a successful anti-islanding detection scheme.

The methods not resident in the inverter are generally controlled by the utility or have communication between the inverter and the utility to affect an inverter shutdown if necessary. These include the impedance insertion method and power line carrier communications.

6.3.4.9 Impedance Insertion

This is also called the resistance insertion method or reactance insertion method as shown in Figure 6.6. The switch is normally open (NO). When the switch opens to disconnect the utility, the capacitor switch is normally closed (NC) with a short delay. If the local delay is of the type that prevents

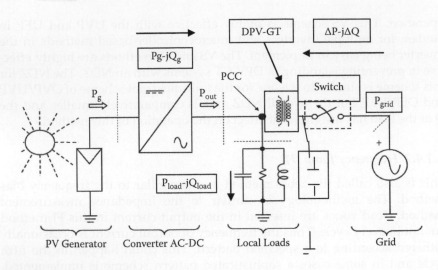

FIGURE 6.6
Resistance insertion method with a capacitor bank with a switch inserted as shown on the local load side before the DPV-GT.

the detection of islanding, then a large capacitor that is switched on in the circuit will cause a shift in balance between the utility and the DPV-GT, causing a jump in the phase that initiates a jump in ω_{res} leading to a detection of islanding by the UFP scheme. Use of a large resistance instead can lead to a jump in the amplitude of the voltage at PCC, thus initiating the OVP scheme. This system is easily implementable with the only weakness that the capacitor banks if not available can add to the overall cost of implementation of such a scheme. However, many such capacitor banks are already available for VAR correction of pf. A minimum NDZ change has to be closely monitored for such a scheme for DPV-GT systems.

6.3.4.10 Power Line Carrier Communications (PLCC)

Tripping generation automatically under loss of communication, where any loss of communication (for 10 seconds) on singular transfer trip signals (no redundancy), requires immediate tripping of generation (i.e., cannot guarantee anti-islanding under loss of communication).

PLCC prevents islanding by sending a low-level signal from the utility side through a transmitter (T) to the local side where it is received by a receiver (R) on the local load side beyond the DPV-GT as shown in Figure 6.7. If the utility gets disconnected then the R does not detect the signal sent by T, an alarm can be sounded to alert the operator of islanding and thus engage the inverters to disconnect. This can be manually initiated by disconnecting the PV inverter and the load separately with their own switches. With increased penetration of DPV-GT systems, this method has increased applications with no NDZ.

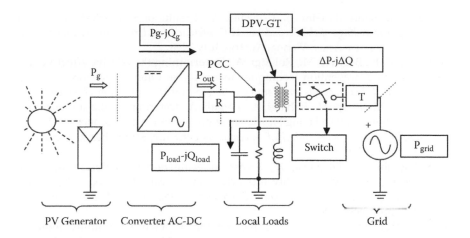

FIGURE 6.7
PLCC with transmitter (T) on the utility side and receiver (R) on the local load side.

6.3.4.11 Supervisory Control and Data Acquisition (SCADA)

This is a logical choice for the prevention of islanding. Most of the utilities use instrumentation from the highest voltage level for controls, so DPV-GT systems are not an exception to the use of SCADA and its utility to utilities. A PV inverter has related sensing devices for voltage monitoring. Thus, when voltage variations occur the OVP/UVP system is already in place to sense islanding issues. In some cases a recloser switch can be used in coordination with an inverter so that out-of phase reclosures do not occur. These phase changes can thus be coordinated with the OFP/UFP system to facilitate further anti-islanding actions. In addition, with SCADA in the system the utility can exercise far better control on the prevention of islanding both in the time-domain and frequency-domain operations. If all necessary instruments are adequately provided, then SCADA should provide easier and better monitoring and control to prevent islanding of DPV-GT grid tied systems. The addition of SCADA at the substation level where DPV-GT systems exist is a complicated operation with multiple inverters if connected for higher-power capacity. However, if properly implemented this system does not have any drawbacks on NDZ.

Bibliography

1. Bower, Ward, and Ropp, Michael, Evaluation of islanding detection methods for utility-interactive inverters in photovoltaic systems, Sandia Report, SAND2002-3591 unlimited release, November 2002.

2. Kern, G., Bonn, R., Ginn, J., Gonzalez, S., Results of SNL grid-tied inverter testing, *Proceedings of the Second World Conference and Exhibition on Photovoltaic Solar Energy Conversion*, Vienna, Austria, July 1998.
3. Ropp, M. E., Begovic, M., Rohatgi, A., Determining the relative effectiveness of islanding prevention techniques using phase criteria and non-detection zones, *IEEE Transactions on Energy Conversion* 15(3), September 2000, 290–296.
4. Begovic, M., Ropp, M., Rohatgi, A., Pregelj, A., Determining the sufficiency of standard protective relaying for islanding prevention in grid-connected PV systems, *Proceedings of the 26th IEEE Specialists Conference*, September 30–October 3, 1997, pp. 1297–1300.
5. Kobayashi, H., Takigawa, K., Statistical evaluation of optimum islanding preventing method for utility interactive small scale dispersed PV systems, *Proceedings of the First IEEE World Conference on Photovoltaic Energy Conversion (1994)*, pp. 1085–1088.
6. IEEE Standard 929-2000, IEEE recommended practice for utility interface of photovoltaic (PV) systems, Sponsored by IEEE Standards Coordinating Committee 21 on Photovoltaics, IEEE Standard 929-2000, Published by the IEEE, New York, April 2000.
7. Jones, R., Sims, T., Imece, A., Investigation of potential islanding of dispersed photovoltaic systems, Sandia National Laboratories report SAND87-7027, Sandia National Laboratories, Albuquerque, NM, 1988.
8. Kobayashi, H., Takigawa, K., Hashimoto, E., Method for preventing islanding phenomenon on utility grid with a number of small scale PV systems, *Proceedings of the 21st IEEE Photovoltaic Specialists Conference (1991)*, pp. 695–700.
9. Best, R., *Phase-locked loops: Theory, design, and applications*, 2nd. ed., McGraw-Hill, New York, 1993.
10. Brennan, P. V., *Phase-locked loops: Principles and practice*, Macmillan, New York, 1996.
11. Ranade, S. J., Prasad, N. R., Omick, S., Kazda, L. F., A study of islanding in utility-connected residential photovoltaic systems, Part I: Models and analytical methods, *IEEE Transactions on Energy Conversion* 4(3), September 1989, 436–445.
12. Wills, R. H., The interconnection of photovoltaic systems with the utility grid: An overview for utility engineers, Sandia National Laboratories Photovoltaic Design Assistance Center, publication number SAND94-1057, October 1994.
13. Handran, D., Bass, R., Lambert, F., Kennedy, J., Simulation of distribution feeders and charger installation for the Georgia Tech Olympic electric tram system, *Proceedings of the Fifth IEEE Workshop on Computers in Power Electronics*, August 11–14, 1996, pp. 168–175.
14. Grebe, T. E., Application of distribution system capacitor banks and their impact on power quality, *IEEE Transactions on Industry Applications* 32(3), May/June 1996, pp. 714–719.
15. Becker, H., Gerhold, V., Ortjohann, E., Voges, B., Entwicklung, aufbau und erste testerfahrung mit einer prüfeinrichtung zum test der automatischen netzüberwachung bei netzgekoppelten wechselrichtern. *Progress in Photovoltaics*.
16. Yuyama, S., Ichinose, T., Kimoto, K., Itami, T., Ambo, T., Okado, C., Nakajima, K., Hojo, S., Shinohara, H., Ioka, S., Kuniyoshi, M., A high-speed frequency shift method as a protection for islanding phenomena of utility interactive PV systems, *Solar Energy Materials and Solar Cells* 35, 1994, 477–486.

17. Ropp, M., Begovic, M., Rohatgi, A., Prevention of islanding in grid-connected photovoltaic systems, *Progress in Photovoltaics* 7, 1999, 39–59.
18. Wyote, A., Belmans, R., Leuven, K., Nijs, J., Islanding of grid-connected module inverters, *Proceedings of the 28th IEEE Photovoltaic Specialists Conference,* September 17–22, 2000, pp. 1683–1686.
19. Chakravarthy, S. K., Nayar, C. V., Determining the frequency characteristics of power networks using ATP, *Electric Machines and Power Systems* 25(4), May 1997, 341–353.
20. Stevens, J., Bonn, R., Ginn, J., Gonzalez, S., Kern, G., Development and testing of an approach to anti-islanding in utility-interconnected photovoltaic systems, Sandia National Laboratories report SAND2000-1939, Albuquerque, NM, August 2000.
21. Kitamura, A., Okamoto, M., Yamamoto, F., Nakaji, K., Matsuda, H., Hotta, K., Islanding phenomenon elimination study at Rokko test center, *Proceedings of the First World Conference on Photovoltaic Energy Conversion, 1994,* pt. 1, pp. 759–762.
22. Kitamura, A., Okamoto, M., Hotta, K., Takigawa, K., Kobayashi, H., Ariga, Y., Islanding prevention measures: demonstration testing at Rokko test center for advanced energy systems, *Proceedings of the 23rd IEEE Photovoltaic Specialists Conference,* 1993, pp. 1063–1067.
23. Toggweiler, P., Summary and conclusions, *Proceedings of the IEA-PVPS Task V Workshop "Grid Connected Photovoltaic System,"* September 15–16, 1997, pp. 15–17.
24. Ropp, M., Aaker, K., Haigh, J., Sabbah, N., Using power line carrier communications to prevent islanding, *Proceedings of the 28th IEEE Photovoltaic Specialists Conference,* September 17–22, 2000, pp. 1675–1678.
25. Riley, C., Lin, B., Habetler, T., Kilman, G., Stator current harmonics and their causal vibrations: A preliminary investigation of sensorless vibration monitoring applications, *IEEE Transactions on Industry Applications* 35(1), January–February 1999, 94–99.
26. Ropp, M., Bonn, R., Gonzalez, S., Whitaker, C., Investigation of the impact of single-phase induction machines in islanded loads—Summary of results, Sandia National Laboratories SAND Report, Albuquerque, NM, 2002.
27. UK Engineering Recommendation G83/1 September 2003, Recommendations for the connection of small-scale embedded generators (up to 16 Amps per phase) in parallel with public low-voltage distribution networks.
28. DISPOWER project (Contract No. ENK5-CT-2001-00522), State-of-the-art solutions and new concepts for islanding protection, 2006 (http://www.pvupscale.org).
29. IEA PVPS Task V, report IEA-PVPVS T5-07: 2002, Probability of islanding in utility networks due to grid connected photovoltaic systems.
30. IEA PVPS Task V, report IEA-PVPS T5-09: 2002, Evaluation of islanding detection methods for photovoltaic utility-interactive power systems.
31. Woyte, A., De Brabandere, K., Van Dommelen, D. l, Belmans, R., Nijs, J., International harmonization of grid connection guidelines: Adequate requirements for the prevention of unintentional islanding, *Progress in Photovoltaics: Research and Applications* 11, 2003, 407–424.
32. Köln, K., Grabitz, A., Kremer, P., Kress, B., Five years of ENS (MSD) islanding protection—What could be the next steps?, *Proceedings of the 17th European Photovoltaic Solar Energy Conference and Exhibition,* Munich, 2001.

33. IEA PVPS Task V, report IEA-PVPS T5-08: 2002, Risk analysis of islanding of photovoltaic power systems within low voltage distribution networks.
34. IEC 61508: Functional safety of electrical/electronic/programmable electronic safety-related systems, 1998.
35. prEN 50438: Requirements for the connection of micro-generators in parallel with public low voltage distribution networks (CENELEC Final Draft), 2007.

7

Relay Protection for Distributed Photovoltaic Grid Power Transformers

7.1 Distributed Photovoltaic Grid Transformer (DPV-GT) Protection

The addition of the DPV energy resource to the distribution or subdistribution level impacts the relay system beyond the point of common coupling (PCC). Generally protection systems are designed for the traditional radial circuit formation. Additional concern is attributed to the bidirectional power flows, increased fault levels, safety, voltage swells, fluctuations and transients, equipment ratings, and autoreclosing. These affect the relay protection schemes applied to the configuration. A distributed resource like DPV when added to the system necessitates the insurance of safe and reliable operation of this distributed resource (DR) like a DPV system.

Presently, most of the distribution systems worldwide operate in the radial configuration, and the power flows in only one direction, especially in the United States and Europe. Installation of DPVs will not alter the topology of the system, but the power will flow in multiple directions. The biggest impact of this is on the protection of distribution systems. Present protection schemes are simple in which fuses are used (as illustrated in Problem 1 at the end of the chapter) for protection of laterals, and the fuses are backed by reclosers on the main feeder or the breaker at the substation. Such simple schemes will not always work with DPVs. Advanced protection schemes, which can adapt to the changing distribution system configuration, are essential. These will depend on the measurement of data at strategic locations and communication of these data to intelligent relays for protection of the system. Thus, protection will become an integral part of distribution automation. Large numbers of DPVs could also lead to stability and frequency control problems. The problems that were only relevant to transmission systems will become relevant to distribution systems. too. Therefore, new technologies to operate and manage the microgrid at the distribution system will be needed.

DPV-GTs are used for a wide variety of applications. The type of protection that will be provided for a DPV-GT depends upon its kilo volt amp (KVA) rating and its importance. The only protection that will possibly be provided to a small lighting transformer may be in the form of fuses, while DPV-GTs connected in a 33 KV station may need elaborate protection.

DPV system protection may be affected due to the following:

1. Distribution systems designed for radial current flow in one direction and one-direction sensing
2. Loss of ground connection in wye-connected DPV-GTs
3. Addition of a DPV system in an existing radially formatted distribution system
4. Change in voltage control under over-voltage protection (OVP)/under-voltage protection (UVP) conditions
5. Islanding conditions
6. Revision of existing autoreclosing schemes
7. Breaker failures and timings due to the addition of DPV-GTs
8. Stability of the entire distribution system

Tables 7.1 and 7.2 summarize the transformer protection scenario and application of various protection schemes.

For faults originating in the transformer, the approximate proportion of faults due to each of the causes listed above is shown in Table 7.2. Thus the need for the OVP/UVP and OVF/UFP is necessary for reliable operation of the DPV scheme.

TABLE 7.1

Internal and External Faults Acting on DPV-GTs

Internal Faults	External Faults
Phase fault	System phase faults
Ground fault	System ground faults
Inter-turn faults	Magnetic inrush currents due to switching loads
Overload: OLTC failure	Over-voltage
Over-fluxing	Under-frequency
Leakage of oil from tank	DC bias or offset
Tank and accessories	Harmonics

TABLE 7.2

Fault Segregation Percentage

Under-frequency	33%
DC bias or offset	10%
Harmonics	33%
Over-voltage	20%
Magnetic inrush currents due to switching loads	4%

7.2 Application of Protective Scheme

7.2.1 Fault Primary Backup

The following are used for backup: phase fault, percentage differential relay, over-current/distance, ground fault, percentage differential relay, over-current/distance, inter-turn fault, Buchholz relay, and oil leaks.

7.2.2 Monitoring True Load

Substation metering schemes must be designed to provide monitoring of true loading on circuits and transformers (planning visibility).

Overload is not allowed on equipment. Special protection (operating) tripping schemes are not allowed, which could provide more flexible (hourly) transformer loading (overload tripping schemes are historically never utilized on transformers).

7.2.3 Direct Transfer Trip (DTT) Communication Requirements

DTT is required when the proposed generation cannot detect a ground or phase fault on the line section when separated from the grid system within 1.5 to 2.0 seconds, or cannot detect an island condition and trip within 2.0 seconds.

7.3 Protection Relays

Some compact multi-type relays that contain the necessary elements for the protection of extra-high-voltage/high-voltage power distribution network facilities and devices such as transformers, generators, and motors, are available in the market and are shown in Figures 7.1 through 7.4. These can be connected to a supervisory monitoring system through the application of the communication function (CC-Link).

The different relays are categorized as follows:

1. *Over-current relay (OCR)*: OCRs from single- to three-phase types are available in the market, with the three-phase type including earth-fault protection. These are suitable for over-current protection on various grounding systems of power networks.
2. *Voltage relay*: These units containing the necessary elements for under-voltage, over-voltage, and earth-fault over-voltage protection are also available in the market and are also suitable for distribution bus bar protection, etc.

FIGURE 7.1
(See color insert) Protection relay for a fault on circuit adjacent to a DPV circuit.

FIGURE 7.2
(See color insert) Voltage relay for a fault on circuit adjacent to a DPV circuit.

3. *Feeder protection relay*: The product lineup offers units containing the necessary over-current and ground-fault protection for the feeder protection of a nongrounded power system. Feeder protection is possible with one unit.
4. *Biased differential relay for transformer protection*: Biased differential relay for transformer protection. Suitable for the connection of various types of transformers.

7.4 Photovoltaic System Ground-Fault Protection

When a photovoltaic system is mounted on the roof of a residential dwelling, National Electric Code (NEC) requirements dictate the installation of ground-fault protection (detection and interrupting) devices (GFPDs).

FIGURE 7.3
(See color insert) Feeder protection relay for a fault on circuit adjacent to a DPV circuit.

FIGURE 7.4
(See color insert) Biased differential relay for a fault on circuit adjacent to a DPV circuit.

However, ground-mounted systems are not required to have the same protection because most grid-connected system inverters incorporate the required GFPDs. Ground-fault detection and interruption circuitry perform ground-fault current detection, fault current isolation, and solar power load isolation by shutting down the inverter. This technology is currently going through a developmental process, and it is expected to become a mandatory requirement in future installations.

7.4.1 Islanding Considerations

With the separation of a utility substation with a DPV-GT tied to the grid, a station may successfully island. Islanding is not encouraged unless it is purposefully conducted. In such cases the following relay schemes need to be included:

1. Numerical relays using alternate setting groups for islanded operation
2. Under-voltage protection which can provide time-delayed protection if conventional over-current protection will not operate
3. Resynchronizing the islanded system

7.4.2 Relay, Fuse, and Line Closer Methodology for Protection of DPV

A DPV system addition to a given distribution system may need timing coordination as well as inclusion of directional current relays in the overall circuit design, because a fault in the existing distribution system can affect the operation of relays in the parallel DPV system due to either fast or slow relays and reclosers. This timing coordination is an important factor to consider because there will be unnecessary interruption of service to customers which otherwise would not have caused the loss of service. In Problem 1 at the end of this chapter, this arrangement is described to illustrate this effect.

7.4.3 Impact on Fuse Saving Schemes

DPV systems in a distribution scheme can affect fuse saving schemes with merely a fuse saving or with a recloser at the breaker in tandem or by using a line recloser. Reclosing helps to clear temporary faults, like a tree touching the lines, without permanent interruption in service. In such case overhead fuses are located for additional protection or sectionalizing. Such fuses are saved on the circuit for temporary faults by de-energizing the line with an upstream de-energizing device before the fuse has a chance to be damaged. Then the reclosing device recloses restoring power beyond the fuse. However, the addition of a DPV system can affect the timing coordination between the reclosing device and the fuse due to the addition of the fault current contributed by the DPV system. An example to illustrate this mechanism is shown in Problem 2 at the end of this chapter.

Bibliography

1. Impact of distributed resources on distribution relay protection, A report to the Line Protection Subcommittee of the Power System Relay Committee of the IEEE Power Engineering Society, prepared by working group D3, April 2004.
2. Venkata, S. S., Pahwa, A., Brown, R. E., Christie, R. D., What future distribution engineers need to learn, *IEEE Transactions on Power Systems* 19(1), February 2004, 17–23.

3. Chao, X. H., System studies for DG projects under development in the US, summary of the panel discussion, IEEE Summer Power Meeting, Vancouver, BC, Canada, 2001.
4. Mozina, C. J., Interconnect protection of dispersed generators, *Proceedings of the Georgia Tech Relay Conference*, May 1999.
5. IEEE Standard 1547-2003, IEEE standard for interconnecting distributed resources with electric power systems.
6. Pettigrew, B., Interconnection of a "Green Power" DG to the distribution system, a case study, *Proceedings of the Georgia Tech Relay Conference*, May 2003.
7. ANSI C84.1. Electric Power Systems and Equipment—Voltage ratings (60 Hz).

Chapter 7 Problems

1. Consider the circuit shown in Figure 7.5. A three-phase ground fault occurs on distribution circuit 1 as shown. Find the contribution from the system of 750 Amps that is validated by the diagram. Suggest a proper choice of recloser and circuit breakers to avoid a disconnect of the DPV circuits.

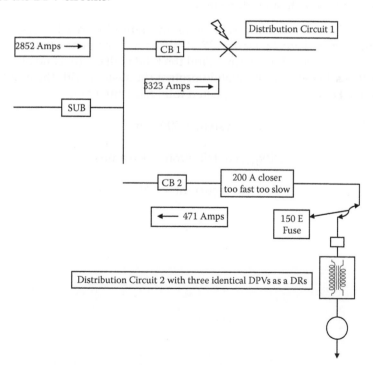

FIGURE 7.5
Fault on circuit adjacent to a DPV circuit.

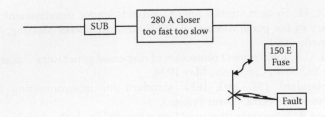

FIGURE 7.6
Fuse-saving example.

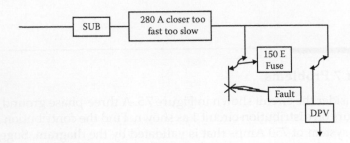

FIGURE 7.7
Fuse-saving example with DPV system added.

2. Consider a fuse-saving scheme as in Figure 7.6. A DPV system is added as shown in Figure 7.7. Assume a three-phase fault on the lateral with fused 150 E fuse. Find the total fault current and evaluate if this scheme will adequately protect the system with the addition of the DPV system. The expected total fault current is

$$I_{fault} \text{ system} = 3000 \text{ amps}$$

$$DPV_{fault} \text{ contribution} = 300 \text{ amps}$$

$$\text{Total fault current} = 3300 \text{ amps}$$

8

DC Bias in Distributed Photovoltaic Grid Power Transformers

8.1 DC Injection into the Grid

Many DC-AC inverters employ transformers in their overall design, thus eliminating the DC bias that can be injected into the grid. Distributed photovoltaic grid transformers (DPV-GTs) play an important role in performing this basic function. However, many commercial inverters are designed without transformers for higher efficiency, lower cost, and size limitations. In such cases a better approach is to design a filter to remove the DC bias.

DC injection into the grid results from badly designed grid-connected inverters. Transformers and inverters can be specially designed to eliminate DC injection into the AC grid. High DC injection causing possible saturation may result in the tripping of a transformer or reduce its life and efficiency, triggering metering errors and heating and burning of cables. There are generally two types of transformers in an inverter to provide galvanic isolation: (1) low-frequency transformers (LFTs) and (2) high-frequency transformers (HFTs). Even harmonics are created in the network by loads and devices that exhibit asymmetrical *i-v* characteristics as shown in Figure 8.1.

Possible sources of even harmonics are three-phase half-wave controlled bridges, AC arc furnaces, converters like three-phase rectifiers supplying DC/DC converters, six-pulse cycloconverters, and half-wave rectifiers.

DC currents in networks need to be classified as

1. Symmetrical DC currents that are created by circuits like half-wave rectifiers and are used in light dimmers and in high-frequency light ballasts for fluorescent lights
2. Asymmetrical DC currents that are created by earth faults in distributed generator (DG) systems using DC sources without galvanic separation or by earth leakage currents in inverter circuits

A large, heavy 50 Hz, LFT prevents DC injection toward the AC side in addition to providing galvanic isolation. Various standards and guidelines define

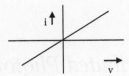

FIGURE 8.1
Asymmetrical *i-v* characteristics of loads and devices, where *i(v)* is not equal to *−i(−v)*.

the maximum DC component that can be tolerated in the grid supply from an standalone photo voltaic (SPV) system.

The DC injection is limited to 0.5% (United States, Australia, and Italy), 1% (Europe, Thailand, and Belgium), and 1 Amp (Germany). Since most Indian standards published by BIS are aligned to IEC standards, DC injection up to 1% is being proposed by the BIS in the Indian standard, keeping with IEC 61727.

DC offset is usually undesirable when it causes saturation or change in the operating point of an amplifier. An electrical DC bias will not pass through a transformer; thus a simple isolation transformer can be used to block or remove it, leaving only the AC component on the other side. In signal processing terms, DC offset can be reduced in real time by a high-pass filter. When one already has the entire waveform, subtracting the mean amplitude from each sample will remove the offset. Often, very low frequencies are called *slowly changing DC* or *baseline wander*.

Pulse-width modulators in inverter designs produce DC bias when even-power harmonics are contained in the basic voltage waveform and in general with a positive-negative imbalance in the reference voltage. This distortion of the reference voltage is especially predominant when the synchronization of the inverter with the network voltage is achieved by using the zero-crossings of the reference voltage waveform.

In extremely cold conditions where the temperature can fall below zero, the grid and solar ties can fail and cause a DC over-voltage condition. In cases where the DC bus exceeds designated maximum voltage say for a system of 600 volts, the inverter has to be shut down and the solar array will be disconnected from the DC bus to protect auxiliary equipment like capacitors, disconnects, and wire insulations at specified nominal voltage.

Such DC bias severely affects DPV-GTs and related current transformers used for metering purposes. Harmonics increase, losses increase, noise levels increase, and half-cycle saturation is seen that can lead to high primary currents and high levels of magnetic inrush currents.

The DC content in the system signal that affects DPV-GTs depends on the different converter topologies that are used, parameters of the PWM pattern, like frequency of modulation and amplitude of modulation, besides the harmonics prevalent in the grid system signal. Experimental results have revealed that the frequency of modulation index (FOMI) directly controls the level of DC. As this FOMI increases, the DC decreases. While there is no direct correlation between the amplitude of modulation index (AOMI) and DC component, the same is the effect of the angle of the harmonic in the system grid signal. The level of the converter plays a significant role in allocating a DC component within the resulting signal. The lower level (i.e., like a two-level converter) generally contributes a higher level of the DC component. Typically for a 500 KVA, 11/0.433 kV distribution transformer with a 50% loading at unity power factor (upf), a generalized typical upper limit of DC injection limit is set at 40 mA DC, with a limit on total harmonic distortion (THD) content of 5%. Lower limits can be set with a better lower limit on the total harmonic distortion.

For different converter topologies like two-level, three-level, or five-level topologies, several parameters of PWM patterns, with modulation frequencies or modulation amplitude indices and harmonics in the grid voltage waveform like even-order 2nd, 4th, 6th, 8th, and 10th, with harmonic percentage of the voltage waveform between 1% and 10% and harmonic angles from zero degrees to 162 degrees, analysis shows that

1. The level of DC voltage is directly proportional to the modulation frequency (MF) index. With the MF index increasing beyond 33 the DC voltage decreases, with the highest level of DC at an MF index of 21.
2. There is no connection between amplitude modulation (AM) index and level of DC voltage present.
3. There is no connection between harmonic angle (HA) and DC voltage.
4. The highest DC levels were created by two-level converters between 0.6% and 9.5%.
5. The DC levels decreased with three-level converters between 1% and 2.5%.
6. The DC levels further decreased with five-level converters between 1.8% and 2.2%.
7. For high-level MF indices (>45), the DC voltages are less than 1%.

8.2 Effects of DC Currents on DPV-GTs

DC injection typically will affect the harmonic distortion of DPV-GTs with transformers loaded up to 50% and upf. A permissible DC injection of up to 5% is allowable. This translates to about 40 mA for a 500 KVA transformer. Another limit of DC injection is about 0.5% of nominal phase current as specified per IEEE 1547. For torroidal transformers up to 3 KVA and

1:1 transformation ratio, a DC current of up to 50% of a nominal rated current does not cause any hazard of a local blackout, as it is averted by the primary fuse of the DPV-GT. In addition, operation under high constant AC current of about 85% of nominal rating and variable DC loads the harmonic contents distortion of the current in the DPV-GT, and this can rise to about 13% without any appreciable damage to the DPV-GT.

Finally, injection of DC into DPV-GT can cause a considerable increase in noise level up to and less than 6 dBa of sound pressure.

Bibliography

1. IEA PVPS Task V, report IEA-PVPS T5-01: 1998, Utility aspects of grid connected photovoltaic power systems.
2. IEA PVPS Task V, report IEA-PVPS V-1-01: 1996, Grid connected photovoltaic power systems: Status of existing guidelines in selected IEA member countries.
3. DISPOWER project (Contract No. ENK5-CT-2001-00522), International standard situation concerning components of distributed power systems and recommendations of supplements, 2005 (http://www.pvupscale.org).
4. DISPOWER project (Contract No. ENK5-CT-2001-00522), Identification of general safety problems, definition of test procedures and design-measures for protection, 2004 (http://www.pvupscale.org).
5. EN 61000-3-2:2006, Electromagnetic compatibility (EMC)—Part 3-2: Limits—Limits for harmonic current emissions (equipment input current ≤16 A per phase).
6. University of Strathclyde, Department of Trade and Industry (Distributed Generation Coordinating Group), DC injection into low voltage AC networks (Contract No. DG/CG/00002/00/00), June 2005 (http://www.dti.gov.uk/publications).
7. National Rural Electric Cooperative Association, Application guide for distributed generation interconnection: 2003 update—The NRECA guide to IEEE 1547, Resource Dynamics Corporation, April 2003.
8. IEEE 1547: 2003, IEEE standard for interconnecting distributed resources with electric power systems.
9. EN 61000-4-13: 2003, Testing and measurement techniques—Harmonics and interharmonics including mains signalling at AC power port, low frequency immunity tests.
10. Hotopp, R., Dietrich, B., Grid perturbations in a housing estate in Germany with 25 photovoltaic roofs, *Proceedings 13th EUPVSEC*, October 1995, Nice, France.
11. IEA-PVPS Task V, report IEA-PVPS T5-2: 1999, Demonstration test results for grid interconnected photovoltaic power systems.
12. IEC 62053-21: 2003, Electricity metering equipment (AC)—Particular requirements—Part 21: Static meters for active energy (classes 1 and 2).
13. Engineering recommendation G83/1 September 2003: Recommendations for the connection of small-scale embedded generators (up to 16 A per phase) in parallel with public low-voltage distribution networks.

9

Thermocycling (Loading) and Its Effects on Distributed Photovoltaic Grid Transformers

In most of the geographical locations in the United States, solar power facilities experience a steady-state loading when inverters are operating. When the sun comes out, there is a dampened reaction process and loading on the transformer is more constant. The entire process is controlled by the insolation number in a particular location. The no-load operation of such transformers is completely controlled by a different set of parameters.

The distributed energy generation from photovoltaic cells is not constant during the day as the incoming solar radiation varies. The presence of intermittent clouds can affect the overall irradiance available at a location. Therefore, an increase in power transport between different regions worldwide is expected. During operation, transformers heat up or cool down depending on the varying loads. This is called thermal cycling or thermocycling. Distributed photovoltaic grid transformers (DPV-GTs) connected to such grids are typically subjected to varying loads. These loads could be linear or nonlinear loads. Nonlinear loads induce total harmonic distortion (THD) currents that can increase losses. These variations in load depend mostly on the time of day, as consumers require more power at specific times of the day. Typically, in North America the load cycles show upswing in the early hours of the day, say from 6:00 AM to 8:00 AM and then from 4:00 PM to 6:30 PM of any day. The high electric field strengths, which are a result of peaking power transport, combined with these thermal cycles have their impact on transformer insulation. The grid of the future will still be partly composed of today's components, and asset management strategies will be used to replace components, such as transformers, just before their end of life. Solar energy as a renewable energy source is often connected to the grid via silicon technology converters like the insulated-gate bipolar transistors (IGBTs) and integrated gate commutated/controlled thyristors (IGCTs). These convertors introduce transient spikes in the grid. Therefore, the effects of both temperature cycling and transients on transformer insulation need to be taken into consideration while designing a transformer connected to a power grid like the DPV-GT.

Solar power systems typically operate very close to their rated loads. Since the load variation from the rated value is appreciably low, the operation of transformers is not adversely affected to cause deterioration of parameters that guide the insulation coordination of the core-coil structure.

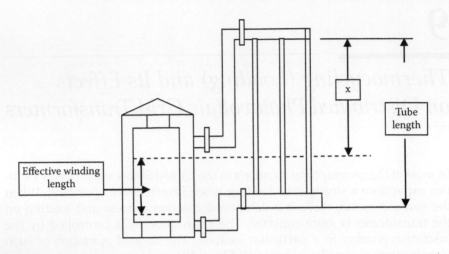

FIGURE 9.1
Transformer tank with radiator tube configuration.

Forces experienced by the primary and secondary windings are not out of the ordinary, thus alleviating problems that may occur in the design of the mechanical structure.

The storage battery interaction with the transformer in a PV system may control the load consistency and alleviate the perceived problems.

Even with the loss correction addressing the harmonic content during the heat run, the hot-spot temperature may not be representative of the real conditions due to the nature of the harmonic current distribution in the winding and how it differs during the heat run. For the DPV-GTs, "extended load run with overload" is recommended by the CIGRE Joint Task Force.

To control thermal cycling for oil-filled DPV-GTs, they are cooled by standard radiators as shown in Figure 9.1. A proper selection of radiators and fans coordinated with the different factors helps to regulate and monitor the thermal cycling effectively.

1. The percentage radiator above effective winding length is given by

$$\% \text{ radiator above effective winding length} = (100 \times)/(\text{tube length}) \quad (9.1)$$

2. The optimum air factor A_f for oil blast (OB) cooling is 2.0, and A_{ff} for oil forced blast (OFB) cooling is 2.81. These factors are generally used in preliminary calculations to determine the fans. After selecting the most suitable fans, then the appropriate air factor is used in calculating the actual OB or OFB dissipation required.

3. The nominal dissipation is given as in Table 9.1. For all these calculations, the tank surface dissipation of 0.55 W/in.² is generally taken into account.

TABLE 9.1

Nominal Dissipation per Radiator

Tube	Number of Tubes per Radiator							
Length	52	56	60	64	68	72	76	80
6'0"	4.54	4.69	5.03	5.37	5.7	6.04	6.38	6.71
6'6"	4.71	5.07	5.44	5.8	6.16	6.52	6.89	7.25
7'0"	5.06	5.45	5.84	6.23	6.62	7.01	7.4	7.79
7'6"	5.41	5.83	6.25	6.66	7.08	7.5	7.91	8.33
8'0"	5.76	6.21	6.65	7.1	7.54	7.98	8.43	8.87
8'6"	6.11	6.58	7.06	7.53	8.0	8.47	8.94	9.41
9'0"	6.46	6.96	7.46	7.96	8.46	8.95	9.45	9.95
9'6"	6.82	7.34	7.87	8.39	8.92	9.44	9.97	10.49
10'0"	7.17	7.72	8.27	8.83	9.37	9.93	10.48	11.03
10'6"	7.52	8.10	8.68	9.26	9.83	10.41	10.99	11.57
11'0"	7.87	8.47	9.08	9.7	10.29	10.90	11.5	12.11
11'6"	8.22	8.85	9.49	10.13	10.75	11.38	12.02	12.65
12'	8.57	9.23	9.89	10.56	11.21	11.87	12.53	13.19

Note: k Watt for 50°C temperature rise (average 0.179 W/in.^{2}).

TABLE 9.2

Correction Factors for Type of Cooling

Type of Cooling	Factors
ON	$H_f \times R_f \times T_f$
OFON	Tank = 1.28
	Radiators
	$1.28 \times R_f \times T_f$
OB	$H_f \times A_f$
OFB	Tank = 1.28
	Radiators
	A_{ff}

9.1 Gradient in Windings with No Directed Oil Flow

Table 9.6 and others in sequence illustrate the factor by which the effective watts/squre inch need to be multiplied to give the gradient above the top oil temperature. This is a necessary step to maintain the proper cooling of the transformer under thermal cycling caused by cloudy skies and intermittent irradiance at a particular site. Besides the top oil surface is the sum total of the net area exposed to the vertical ducts plus the effective net area exposed to the horizontal ducts (Table 9.7).

TABLE 9.3

Correction Factors for Percent Radiator Above Effective Winding

Percent Radiator Above Effective Winding	20	25	30	35	40	45	50	55	60	65	70	75	80	85	90	95
H_f	.781	0.806	0.824	0.839	.855	.871	.887	.903	.919	.935	.951	.967	.984	1.0	1.015	1.032

9.? Gradient in Winding with Multidirected Oil Flow

Table 9.6 and others in respective chapters are a factor by which the effective water/oil gradient has to be multiplied to yield a gradient above the top oil temperature. That is, to ensure sufficient and efficient the proper cooling in that transformer oil it is necessary to ensure a smooth, slow, and controlled distribution in a given the sliding heat of surfaces as the sum total of the radiator exposed to the windings ... vertical and the attached vertical area exposed to the horizontal ducts (Table 9.?)

TABLE 9.4

Fan Selection to Optimize Effects of Thermocycling in DPV-GTs

Fan Diameter (inches)	Free Air Delivery (cu.ft/min)	Noise Level at 40 dB Weighing	Radiators per Fan	Minimum Number of Tubes/Row
30	11,950 @ 900 rpm	70 dB @ 900 rpm	2–3	54
36	16,050 @ 700 rpm	70 dB @ 700 rpm	2–3	64
42	20,300 @ 560 rpm	69 dB @ 560 rpm	3–4	74
48	25,600 @ 560 rpm	69 dB @ 470 rpm	4–5	80

TABLE 9.5

Pipe Sizes and Rating (in kW)

Pipe Diameter (inches)	Radiator kW for ON	Radiator kW for OB
2	9.5	14
3	20	30
4	35	52
5	54	81
6	77	115
7	101	151
8	130	195
9	162	243
10	195	295
11	230	345
12	270	405

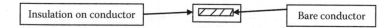

FIGURE 9.2
Bare conductor with insulation covering.

If the radial depth of the coil is unexpectedly exceeded for a given duct width, the additional horizontal area should not be included in estimating the total effective area of the cooling surface.

Finally, the cooling equipment needs to be designed to give the following margins below the guaranteed temperature rise:

Temperature rise in oil between <1 ½°C to 48 ½°C> for a 50°C rise

Temperature rise by resistance between <2°C to 58°C> rise for 60°C rise

Temperature rise by resistance between <3°C to 62°C> rise for 65°C rise

The above analysis helps to adjust the thermal cycling of DPV-GTs caused by cloudy skies.

TABLE 9.6

Factor for Gradient in Windings without Directed Oil Flow for
Disc Winding as in Figure 9.3

Insulation Covering MILS*	Factor for ON and OB	Factor for OFN and OFB
20	11	14.2
30	12.25	15.8
40	13.5	17.5
50	14.75	19.1
60	16	20.7
70	17.25	22.3
80	18.5	24.0
90	19.75	25.6
100	21	27.2
110	22.25	28.8
120	23.5	30.5
130	24.75	32.1
140	26	33.7
150	27.25	35.3
160	28.5	37
170	29.75	38.6
180	31	40.2

* Thousandths of an inch.

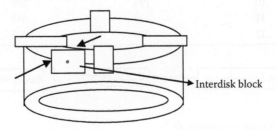

FIGURE 9.3
Disc winding with interdisc blocks and bare conductor with insulation covering.

9.2 Some Commercially Available Epoxy Materials and Their Advantages for DPV-GTs

1. Minimal or virtually no shrinkage, optimum dielectric properties, high dimensional stability, excellent adhesion and thermal cycling properties
2. Two-component systems that harden at room temperature and have service temperatures between 100°C and 130°C
3. One- and two-component heat-cured systems that can resist temperatures up to 200°C

TABLE 9.7

Effective Area for Horizontal Ducts in Windings

Depth (inches)	Percent Effective Area
¼	93
½	87
¾	82
1	76
1 ¼	71
1 ½	67
1 ¾	63
2	60
2 ½	54
3	50
3 ½	47
4	44
4 ½	41.5
5	39.5
5 ½	37.5
6	36

TABLE 9.8

Radial Depth and Axial Length of Windings in Relation to Width of Ducts

Ducts (inches)	Radial Depth (inches)	Axial Length (inches)
3/32	¾–7/8	15
1/8	1 3/8–1 ½	30
5/32	2 3/16–2 3/8	45
3/16	3–3 ¼	60
7/32	3 ¾–4 1/8	75
¼	4 ½–5	90
9/32	5 ½–6	105
5/16	6 3/8–7	120
11/32	7 ¼–8	135
3/8	8 1/8–9	150

4. Filled potting compounds containing inorganic ingredients that impart thermal conductivity, a low coefficient of thermal expansion, low shrinkage, and fire retardancy

Unfilled compounds are mostly transparent and are used when the lowest viscosity is needed and defective components need to be identified easily for replacement and repair.

All solar compounds as fire-retardant (FR) products are fire resistant, and where specified, Underwriter Laboratories recognized as Flame Class UL 94V-0. Custom formulas are available in all categories and are packaged in 1, 5, and 55 gal. kits.

9.3 Some Commercially Available Products and Their Applications

Castings are available of sensitive electronic components, potting telecommunications equipment, thermal cut-out switches, potting ballasts, pumps, surge suppressors, connectors, switches, relays, coils, transformers, power supplies, resistors, solenoids, proximity switches, transistors, sensors, power line filters, timers, etc.

For dry-type or resin-cast DPV transformers, suitable solar potting and encapsulating compounds are used to abate and mitigate thermal cycling effects. Some of these compounds available in the market are described below.

Solar potting and encapsulating compounds extend the life of DPV-GTs, power supplies, electronic devices, and modern high intensity discharge (HID) automotive headlight systems through improved heat dissipation and component protection from shock and vibration.

> EP-0597: Moderate-viscosity unfilled epoxy potting and casting compound; transparent, semirigid cure with Solarcure 5; excellent thermal shock resistance and good electrical properties; with Solarcure-6, yields excellent adhesion and low durometer; low stress and good thermal shock properties that allow component replacement and repair

> EP-211FRHTC_SC1-2: A thermally conductive, moderate-viscosity epoxy potting and casting compound that can be room temperature or heat cured depending upon the hardener used. With Solarcure 1 (SC1), a UL 94V-0 product with a service temperature of up to 130°C can be produced. With Solarcure 2 (SC2), a heat-cure product with a service temperature of up to 155°C can be achieved. Other custom Solarcure hardeners can be used to adjust viscosity, hardness, cure time, exotherm, service temperature, thermal shock and resistance, etc.

> EP-211FRHTC_SC2-2B: A thermally conductive, fire retardant, UL 94V-0 recognized, moderate-viscosity epoxy potting and casting compound; room temperature or heat cured depending upon the hardener chosen; service temperature up to 155°C with Solarcure 2 (SC2); service temperature up to 200°C with Solarcure 2B (SC2B). Other Solarcure hardeners can be used to adjust viscosity, hardness, cure time, exotherm, service temperature, thermal shock resistance, etc.

10

Power Quality Provided by Distributed Photovoltaic Grid Power Transformers

10.1 Power Quality Requirements

Power quality is characterized by stable nominal AC voltage and frequency within specified tolerances, free from voltage flicker, voltage sags or dips, harmonics, frequency variations, phase angle variations etc. With high connection densities of SPV systems to be implemented in distribution networks, low irradiance can lead to poor power quality, causing failure of sensitive electric bulbs, electronic equipment such as TVs, washing machines, ACs, microwaves, computers etc, or reduce their life. So the design of the inverter plays a very important role. It supervises various parameters, controls them and takes corrective action. A good inverter produces a qualitative AC with the following minimum requirements:

- Low harmonics and waveform distortion: the presence of odd and even harmonics is limited to less than a THD of 5%. The negative effects of harmonics on transformers are commonly unnoticed and disregarded until an actual failure happens. Generally, transformers designed to operate at rated frequency have had their loads replaced with nonlinear types, which inject harmonic currents into the system.

- Low DC injection into the grid: schemes involving MPPT systems need to be designed to satisfy the requirements of the necessary IEEE specifications: 519/1547

- Within limit voltage unbalance

- Limited voltage fluctuation

- Limited voltage flicker: the silicon circuitry used in the inverters leads to corruption of the main supply line resulting into voltage flicker that can cause the light bulbs of residential customers affect very severely causing them to fuse over time and get destroyed. Voltage flicker also affects the eye-sight of individuals constantly using computers as the screens can experience a change in the luminance.

- Under or over frequency: Since the frequency variation can come from the network only, therefore no difference to a "standard" power transformer is expected to be made in its design and manufacture. However, the change in frequency can cause a change in the stray losses, thus the design of the DPV-GTs have to take precautionary steps to reduce the effect of increase in such losses by providing adequate aluminum or cold-rolled grain oriented (CRGO) shunts where high currents are likely to flow in the leads and tank material.

The more critical issues are:

- Under and over voltage
- Power factor control

Power quality determines the fitness of electrical power to consumer devices. Synchronization of the voltage frequency and phase allows electrical systems to function in their intended manner without significant loss of performance or life. The term is used to describe electric power that drives an electrical load and the load's ability to function properly. Without the proper power, an electrical device (or load) may malfunction, fail prematurely or not operate at all. There are many ways in which electric power can be of poor quality and many more causes of such poor quality power.

The electric power industry comprises electricity generation (AC power), electric power transmission and ultimately electricity distribution to an electricity meter located at the premises of the end user of the electric power. The electricity then moves through the wiring system of the end user until it reaches the load. The complexity of the system to move electric energy from the point of production to the point of consumption combined with variations in weather, generation, demand and other factors provide many opportunities for the quality of supply to be compromised.

While "power quality" is a convenient term for many, it is the quality of the voltage—rather than power or electric current—that is actually described by the term. Power is simply the flow of energy and the current demanded by a load is largely uncontrollable.

The quality of electrical power may be described as a set of values of parameters, such as:

- Continuity of service
- Variation in voltage magnitude (see below)
- Transient voltages and currents
- Harmonic content in the waveforms for AC power

It is often useful to think of power quality as a compatibility problem: is the equipment connected to the grid compatible with the events on the grid, and is the power delivered by the grid, including the events, compatible with the equipment that is connected? Compatibility problems always have at least two solutions: in this case, either clean up the power, or make the equipment tougher.

The tolerance of data-processing equipment to voltage variations is often characterized by the CBEMA curve, which give the duration and magnitude of voltage variations that can be tolerated [1].

Ideally, AC voltage is supplied by a utility as sinusoidal having an amplitude and frequency given by national standards (in the case of mains) or system specifications (in the case of a power feed not directly attached to the mains) with an impedance of zero ohms at all frequencies.

No real-life power source is ideal and generally can deviate in at least the following ways:

- Variations in the peak or RMS voltage are both important to different types of equipment.
- When the RMS voltage exceeds the nominal voltage by 10 to 80% for 0.5 cycle to 1 minute, the event is called a "swell."
- A "dip" (in British English) or a "sag" (in American English the two terms are equivalent) is the opposite situation: the RMS voltage is below the nominal voltage by 10 to 90% for 0.5 cycle to 1 minute.
- Random or repetitive variations in the RMS voltage between 90 and 110% of nominal can produce a phenomenon known as "flicker" in lighting equipment. Flicker is rapid visible changes of light level. Definition of the characteristics of voltage fluctuations that produce objectionable light flicker has been the subject of ongoing research.
- Abrupt, very brief increases in voltage, called "spikes," "impulses," or "surges," generally caused by large inductive loads being turned off, or more severely by lightning.
- "Undervoltage" occurs when the nominal voltage drops below 90% for more than 1 minute. The term "brownout" is an apt description for voltage drops somewhere between full power (bright lights) and a blackout (no power – no light). It comes from the noticeable to significant dimming of regular incandescent lights, during system faults or overloading etc., when insufficient power is available to achieve full brightness in (usually) domestic lighting. This term is in common usage has no formal definition but is commonly used to describe a reduction in system voltage by the utility or system operator to decrease demand or to increase system operating margins.
- "Overvoltage" occurs when the nominal voltage rises above 110% for more than 1 minute.
- Variations in the frequency.
- Variations in the wave shape—usually described as harmonics.

- Nonzero low-frequency impedance (when a load draws more power, the voltage drops).
- Nonzero high-frequency impedance (when a load demands a large amount of current, then stops demanding it suddenly, there will be a dip or spike in the voltage due to the inductances in the power supply line).

Each of these power quality problems has a different cause. Some problems are a result of the shared infrastructure. For example, a fault on the network may cause a dip that will affect some customers; the higher the level of the fault, the greater the number affected. A problem on one customer's site may cause a transient that affects all other customers on the same subsystem. Problems, such as harmonics, arise within the customer's own installation and may propagate onto the network and affect other customers. Harmonic problems can be dealt with by a combination of good design practice and well proven reduction equipment.

10.1.1 Power Conditioning

Power conditioning is modifying the power to improve its quality.

An uninterruptible power supply can be used to switch off of mains power if there is a transient (temporary) condition on the line. However, cheaper UPS units create poor-quality power themselves, akin to imposing a higher-frequency and lower-amplitude square wave atop the sine wave. High-quality UPS units utilize a double conversion topology which breaks down incoming AC power into DC, charges the batteries, then remanufactures an AC sine wave. This remanufactured sine wave is of higher quality than the original AC power feed [2].

A surge protector or simple capacitor or varistor can protect against most overvoltage conditions, while a lightning arrestor protects against severe spikes.

Electronic filters can remove harmonics.

10.1.2 Smart Grids and Power Quality

Modern systems use sensors Electronic filters called phasor measurement units (PMU) distributed throughout their network to monitor power quality and in some cases respond automatically to them. Using such smart grids features of rapid sensing and automated self healing of anomalies in the network promises to bring higher quality power and less downtime while simultaneously supporting power from intermittent power sources and distributed generation, which would if unchecked degrade power quality.

A Power Quality Compression Algorithm is an algorithm used in power quality analysis. To provide high quality electric power service, it is essential to monitor the quality of the electric signals also termed as power quality (PQ) at different locations along an electrical power network. Electrical utilities carefully monitor waveforms and currents at various network locations constantly, to understand what lead up to any unforeseen events such as a power outage and blackouts. This is particularly critical at sites where the environment and public safety are at risk (institutions such as hospitals, sewage treatment plants, mines, etc.).

10.1.3 Power Quality Challenges

Engineers have at their disposal many meters [1] that are able to read and display electrical power waveforms and calculating parameters of the waveforms. These parameters may include, for example, current and voltage RMS, phase relationship between waveforms of a multi-phase signal, power factor, frequency, THD, active power (KWatt), reactive power (KVAR), apparent power (KVA) and active energy (KWh), reactive energy (KVARH) and apparent energy (KVAh) and many more. In order to sufficiently monitor unforeseen events, Ribeiro et al. [2] explain that it is not enough to display these parameters, but also to capture voltage waveform data at all times. This is impracticable due to the large amount of data involved, causing what is known as the "bottle effect." For instance, at a sampling rate of 32 samples per cycle, 1,920 samples are collected per second. For three-phase meters that measure both voltage and current waveforms, the data is 6 to 8 times as much. More practical solutions developed in recent years store data only when an event occurs (for example, when high levels of power system harmonics are detected) or alternatively to store the RMS value of the electrical signals [3]. This data, however, is not always sufficient to determine the exact nature of problems.

10.1.4 Raw Data Compression

Nisenblat et al. [4] proposes the idea of power quality compression algorithm (similar to the Lossy Compression method) that enables meters to continuously store the waveform of one or more power signals, regardless whether or not an event of interest was identified. This algorithm referred to as PQZip empowers a processor with a memory that is sufficient to store the waveform, under normal power conditions, over a long period of time, of at least a month, two months or even a year. The compression is performed in real time, as the signals are acquired; it calculates a compression decision before all the compressed data is received. For instance should one parameter remain constant, and various others fluctuate, the compression decision retains only what is relevant from the constant data, and retains all the fluctuation data. It then decomposes the waveform of the power signal of numerous components, over various periods of the waveform. It concludes the process by compressing the values of at least some of these components over different periods, separately. This real time compression algorithm, performed independent of the sampling, prevents data gaps and has a typical 1000:1 compression ratio.

10.1.5 Aggregated Data Compression

A typical function of a common power quality analyzer is a generation of data archive aggregated over given interval. Most typically 10 minutes or 1 minute interval is used as specified by the IEC/IEEE PQ standards. A significant archive sizes are created during an operation of such instrument. As Kraus et. al. [5] have demonstrated the compression ratio on such archives using Lempel–Ziv–Markov chain algorithm, bzip or other similar loss less compression algorithms can be significant. By using prediction and modeling on the stored time series in the actual power quality archive the efficiency of post processing compression is usually further improved. This combination of simplistic techniques implies savings in both data storage and data acquisition processes.

10.2 Power Quality in Grid Connected Renewable Energy Systems

Centralized power generation systems are facing the twin constraints of shortage of fossil fuel and the need to reduce emissions. Long transmission lines are one of the main causes for electrical power losses. Therefore, emphasis has increased on distributed generation (DG) networks with integration of renewable energy systems into the grid, which lead to energy efficiency and reduction in emissions. With the increase of the renewable energy penetration to the grid, power quality (PQ) of the medium to low voltage power transmission system is becoming a major area of interest. Most of the integration of renewable energy systems to the grid takes place with the aid of power electronics converters. The main purpose of the power electronic converters is to integrate the DG to the grid in compliance with power quality standards. However, high frequency switching of inverters can inject additional harmonics to the systems, creating major PQ problems if not implemented properly. Custom Power Devices (CPD) like STATCOM (Shunt Active Power Filter), DVR (Series Active Power Filter) and UPQC (Combination of series and shunt Active Power Filter) are the latest development of interfacing devices between distribution supply (grid) and consumer appliances to overcome voltage/current disturbances and improve the power quality by compensating the reactive and harmonic power generated or absorbed by the load. Solar and wind are the most promising DG sources and their penetration level to the grid is also on the rise. Although the benefits of DG includes voltage support, diversification of power sources, reduction in transmission and distribution losses and improved reliability [1], power quality problems are also of growing concern. This paper deals with a technical survey on the research and development of PQ problems related to solar and wind energy integrated to the grid and the impact of poor PQ. The probable connection topologies of CPDs into the system to overcome the PQ problems are also discussed. A custom power park concept for the future grid connection of distributed generation system is mentioned.

10.3 Power Quality Issues (DG)

Approximately 70 to 80% of all power quality related problems can be attributed to faulty connections and/or wiring [2]. Power frequency disturbances, electromagnetic interference, transients, harmonics and low power factor are the other categories of PQ problems (shown in Table 10.1) that are related to the source of supply and types of load [3].

Among these events, harmonics are the most dominant one. The effects of harmonics on PQ are specially described in [4]. According to the IEEE standard, harmonics in the power system should be limited by two different methods; one is the limit of harmonic current that a user can inject into the utility system at the point of common coupling (PCC) and the other is the limit of harmonic voltage that the utility can supply to any customer at the PCC. Details of these limits can be found in [5]. Again, DG interconnection standards are to be followed considering PQ, protection and stability issues [6].

TABLE 10.1

Categories of Power Quality (PQ) Problems

Power Frequency Disturbance	Low-Frequency Phenomena	Harmonics Even and Odd with PQ Problems
Electromagnetic interferences	Produce voltage sag/swell	Low-frequency phenomena
Power system transient	High-frequency phenomena	Caused by direct current or induced electrostatic field
Power system harmonics	Interaction between electric and magnetic field	Produce waveform distortion
Electrostatic discharge	Fast, short-duration event	Current flow with different potentials
Power factor	Produce distortion like notch, impulse	Low power factor causes equipment damage

10.4 Grid integration of Renewable Energy Systems— Power Quality Issues A Solar Photovoltaic Systems

Though the output of a PV panel depends on the solar intensity and cloud cover, the PQ problems not only depend on irradiation but also are based on the overall performance of solar photovoltaic system including PV modules, inverter, filters controlling mechanism etc. Studies presented in [7], show that the short fluctuation of irradiance and cloud cover play an important role for low-voltage distribution grids with high penetration of PV. Therefore, special attention should be paid to the voltage profile and the power flow on the line. It also suggests that voltage and power mitigation can be achieved using super-capacitors which result in an increase of about 20% in the cost of the PV system.

Voltage swell may also occur when heavy load is removed from the connection. Concerning DG, voltage disturbance can cause the disconnection of inverters from the grid and therefore result in losses of energy. Also, long-term performance of grid connected PV systems shows a remarkable degradation of efficiency due to the variation of source and performance of inverter [8]. Behavior of a very sensitive inverter (fast disconnection for a short and shallow voltage sag) can be found in [9] Fig. 10.2, general structure of grid-connected PV system. The general block diagram of a grid-connected PV system is shown in [9] Fig 10.2 and the system can be a single-phase or three-phase depending on the grid connection requirements. The PV array can be a single or a string of PV panels either in series or parallel mode connection. Centralized or decentralized mode of PV systems can also be used and the overview of these PV-Inverter-Grid connection topologies along with their advantages and disadvantages are discussed in [10].

These power electronics converters, together with the operation of non-linear appliances, inject harmonics to the grid. In addition to the voltage fluctuation due to irradiation, cloud cover or shading effects could make the PV system unstable in terms of grid connection. Therefore, this needs to be considered in the controller design for the inverter [11–12]. In general, a grid-connected PV inverter is not able to control the reactive and harmonic currents drawn from non-linear loads. An interesting controlling mechanism has been presented in [13] where a PV system is used as an active filter to compensate the reactive and harmonic current as well as injecting power to the grid. This system can also operate in stand-alone mode. But the overall control circuit becomes somewhat more complex. Research [14] also shows that remarkable achievements have been made on improving inverter control to provide the reactive power compensation and harmonic suppression as ancillary services. A multifunctional PV Inverter for a grid connected system (Fig 3) has been developed recently and presented in [15]. This system demonstrates reliability improvement through UPS functionality, harmonic compensation, and reactive power compensation capability together with the connection capability during the voltage sag condition. However, the results show that the PQ improvement remains out of the IEEE range. Concept of a Multifunctional PV-Inverter System integrated into an industrial grid can be found in [15].

10.5 Mitigation of PQ Problems

There are two ways to mitigate the power quality problems—either from the customer side or from the utility side. The first approach is called load conditioning, which ensures that the equipment is less sensitive to power disturbances, allowing the operation even under significant voltage distortion. The other solution is to install line conditioning systems that suppress or counteracts the power system disturbances. Several devices including flywheels, super-capacitors, other energy storage systems, constant voltage transformers, noise filters, isolation transformers, transient voltage surge suppressors, harmonic filters are used for the mitigation of specific PQ problems. Custom power devices (CPD) like DSTATCOM, DVR and UPQC are capable of mitigating multiple PQ problems associated with utility distribution and the end user appliances. The following section of the paper looks at the role of CPDs in mitigating PQ problems in relation to grid integrated solar and wind energy systems.

10.6 Role of Custom Power Devices

Voltage quality which ultimately affects Power Quality can be seen as an umbrella name for deviations from ideal voltage conditions at a site in a network. This is equivalent to electromagnetic disturbances of the voltage at the site. With no disturbances the voltage quality is perfect, otherwise it is not. Electromagnetic disturbances are defined as electromagnetic phenomena that may degrade the performance of equipment. Adequate voltage quality contributes to the satisfactory function of electrical and electronic equipment in terms of electromagnetic compatibility. Electromagnetic disturbances as imperfect voltage quality at a site in a network can be regarded as electromagnetic emission from the network, according to the EMC Directive, a network is equipment.

The Custom Power (CP) concept was first introduced by N.G. Hingorani in 1995 [23]. Custom Power embraces a family of power electronic devices, or a toolbox, which is applicable to distribution systems to provide power quality solutions. This technology has been made possible due to the widespread availability of cost effective high power semiconductor devices such as GTOs and IGBTs, low cost microprocessors or microcontrollers and techniques developed in the area of power electronics. DSTATCOM is a shunt-connected custom power device specially designed for power factor correction, current harmonics filtering, and load balancing. It can also be used for voltage regulation at a distribution bus [26]. It is often referred to as a shunt or parallel active power filter. It consists of a voltage or a current source PWM converter ([25] as in Fig. 10.8). It operates as a current controlled voltage source and compensates current harmonics by injecting the harmonic components generated by the load but phase shifted by 180 degrees. With an appropriate control scheme, the DSTATCOM can also compensate for poor load power factor. As in [24], Figure 10.8), the system configuration of DSTATCOM, the DVR is a series-connected custom power device to protect sensitive loads from supply side disturbances (except outages). It can also act as a series active filter, isolating the source from harmonics generated by loads. It consists of a voltage-source PWM converter equipped with a dc capacitor and connected in series with the utility supply voltage through a low pass filter (LPF) and a coupling transformer ([26] as shown in Fig 10.9). This device injects a set of controllable ac voltages in series and in synchronism with the distribution feeder voltages such that the load-side voltage is restored to the desired amplitude and waveform even when the source voltage is unbalanced or distorted. As shown in ([26 Figure 10.9(a) Rectifier supported Figure 10.9(b) DC capacitor supported DVR) UPQC is the integration of series and shunt active filters, connected back-to-back on the dc side and share a common DC capacitor ([27] as shown in Fig 10.10). The series component of the UPQC is responsible for mitigation of the supply side disturbances: voltage sags/swells, flicker, voltage unbalance and harmonics. It inserts voltages so as to maintain the load voltages at a desired level; balanced and distortion free. The shunt component is responsible for mitigating the current quality problems caused by the consumer: poor power factor, load harmonic currents, load unbalance etc. It injects currents in the ac system such that the source currents become balanced sinusoids and in phase with the source voltages. The application of the STATCOM is already reported for wind power applications in stability enhancement, transient, flicker mitigation etc. [28–29]. As the traditional STATCOM works only in leading and lagging operating mode, its application is therefore limited to reactive power support only. The fluctuating power due to the variation of wind cannot be smoothed by using a STATCOM, because it has no active power control ability. To overcome this problem, Battery Energy Storage System (BESS) has been incorporated with STATCOM (STATCOM/BESS) ([30], Fig. 10.10: System configuration of UPQC.

Fig. 10.11 (STATCOM BES and BR) which has both real and reactive power control ability to improve power quality and similarly the DVR can also be used with BESS to control the stability of wind farm [30]. The reactive and active power flow with harmonic voltage mitigation for a grid-connected, distributed generation system [31] is shown in Figure 10.12, power quality control using DVR and BESS. Every recent research report [32, 33] shows that significant structure has been proposed in [32] and in Figure 10.13.

Research and development has been carried out on the application of UPQC to grid-connected PV and wind energy systems. As the UPQC can compensate for almost all existing PQ problems in the transmission and distribution grid, placement of a UPQC in the distributed generation network can be multipurpose.

It works both in interconnected and islanded mode where PV is connected to the DC link in the UPQC as an energy source. UPQC has the ability to inject power using PV to sensitive loads during source voltage interruption. The advantage of this system is voltage interruption compensation and active power injection to the grid in addition to the other normal UPQC abilities. But the system's functionality may be compromised if the solar resource is not sufficient during the voltage interruption condition. UPQC with grid connected PV ([32] as in Figure 10.13) the application of a UPQC to overcome the grid integration problems of the fixed speed induction generator (FSIG) is investigated in ([33], as in Figure 10.14) The FSIG fails to remain connected to the grid in the event of grid voltage dip or line fault due to excessive reactive power requirement.

The drop in voltage creates over speeding of the turbine, which causes a protection trip. With the aid of the UPQC, this fault-ride-through capability is achieved, which greatly enhances system stability. Result show that the UPQC as one of the best devices for the integration of wind energy system to the grid. A grid-connected wind energy system with UPQC [13] and the concept of a custom power park has been proposed using CPDs to provide quality power at various levels. It has been extended further by using supervisory control techniques to coordinate the custom power devices by proving the pre-specified quality of power.

10.7 Effects of irradiance in a Solar Photovoltaic Systems

Though the output of PV panel depends on the solar intensity and cloud cover, the PQ problems are not only depends on irradiation but also based on the overall performance of solar photovoltaic system including PV modules, inverter, filters controlling mechanism etc. Studies, presented in [7], shows that the short fluctuation of irradiance and cloud cover play an important role for the low-voltage distribution grids with high penetration of PV. Therefore special attention should be paid to the voltage profile and the power flow on the line. It also suggests that Voltage and power mitigation can be done using super capacitor which increases around one fifth of the PV system cost. Voltage swell may also occur when heavy load is removed from the connection.

Concerning DG, any kind of voltage disturbance causes the disconnection of inverter from the grid and therefore it results losses of energy. Also long term performance of grid connected PV system shows a remarkable degradation of efficiency due to the variation of source and performance of inverter [8].

The general block diagram of grid connected PV system is shown in [2] and the system can be a single-phase or three phase depending on the grid connection requirements. The PV array can be a single or a string of PV panels either in series or parallel mode connection. Centralized or decentralized mode of PV systems can also be used and the overview of these PV-Inverter-Grid connection topologies along with their advantages and disadvantages are d out in [9].

These power electronics converters along with the operation of non-linear appliances inject harmonics to the grid. On the over hand voltage fluctuation due to the irradiation, cloud cover or shading effect could make the PV system instable in term of grid connection. Therefore, special control design is required in the controlling part of the inverter [10–11].

In general, grid-connected PV inverter is not able to control the reactive and harmonic currents drawn from the non-linear loads. An interesting controlling mechanism has been presented in [12] where PV system is used as an active filter to compensate the reactive and harmonic current as well as injecting power to the grid. This system can also operate in stand alone mode. But the overall control circuit becomes somewhat more complex. Present research [13] also shows that a remarkable achievement has been done on improvement of Inverter control to provide the reactive power compensation and harmonic suppression as ancillary services. A multifunctional PV Inverter for grid connected system [3] has been developed recently and presented in [14] shows the reliability through UPS functionality, harmonic compensation, reactive power compensation capability along with the connection capability during the voltage sag condition. But result shows that PQ improvement is still out of IEEE range.

References

1. Galli et al., Exploring the power of wavelet analysis?: Oct 1996, IEEE, *IEEE Computer Applications in Power*, vol. 9, issue 4, pp. 37–41.
2. Ribeiro et al., An enhanced data compression method for applications in power quality analysis? Nov. 29-Dec. 2, 2001, IEEE, The 27th Annual Conference of the IEEE Industrial Electronics Society, 2001. *IECON '01*, vol. 1, pp. 676–681.
3. Ribeiro et al., An improved method for signal processing and compression in power quality evaluation? Apr. 2004, IEEE, *IEEE Transactions on Power Delivery*, vol. 19, issue 2, pp. 464–471.
4. Nisenblat et al., Method of compressing values of a monitored electrical power signal. April 18, 2004.

5. Kraus, Jan; Tobiska, Tomas; Bubla, Viktor, "Lossless encodings and compression algorithms applied on power quality datasets," Electricity Distribution—Part 1, 2009. CIRED 2009. *20th International Conference and Exhibition*, vol., no., pp. 1–4, 8–11 June.
6. pge.com—A utility pamphlet illustrating the CBEMA curve.
7. datacenterfix.com—A Power Quality discussion on UPS design.
8. Dugan, Roger C.; Mark McGranaghan, Surya Santoso, H. Wayne Beaty (2003). *Electrical Power Systems Quality*. McGraw-Hill Companies, Inc. ISBN 0-07-138622-X.
9. Meier, Alexandra von (2006). *Electric Power Systems: A Conceptual Introduction*. John Wiley & Sons, Inc. ISBN 978-0471178590.
10. Heydt, G.T. (1994). Electric Power Quality. Stars in a Circle Publications. Library Of Congress 621.3191.
11. Bollen, Math H.J. (2000). *Understanding Power Quality Problems: Voltage Sags and Interruptions*. New York: IEEE Press. ISBN 0-7803-4713-7.
12. Sankaran, C. (2002). *Power Quality*. CRC Press LLC. ISBN 0-8493-1040-7.
13. Baggini, A. (2008). *Handbook of Power Quality*. Wiley. ISBN 978-0-470-06561-7.
14. Kusko, Alex; Marc Thompson (2007). *Power Quality in Electrical Systems*. McGraw Hill.
15. IEEE Standard 519 Recommended Practices and Requirements for Harmonic Control in Electrical Power Systems section 10.5 Flicker.
16. I M de Alegría, J Andreu, J L Martín, P Ibanez, J L Villate, H Camblong, "Connection requirements for wind farms: A survey on technical requierements and regulation," *Renewable and Sustainable Energy Reviews*, 2007, vol. 11, 1858–1872.
17. F Blaabjerg, R Teodorescu, M Liserre, A V. Timbus, "Overview of Control and Grid Synchronization for Distributed Power Generation Systems," *IEEE Trns Indust Elect*, 2006, Vol. 53(5), pp. 1398–1409.
18. S M Dehghan, M Mohamadian and A Y Varjani, "A New Variable-Speed Wind Energy Conversion System Using Permanent Magnet Synchronous Generator and Z-Source Inverter," *IEEE Trns Energy Conv*, 2009, Vol 24(3), 714–724.
19. S.P. Chowdhurya, S. Chowdhurya, P.A. Crossley, "Islanding protection of active distribution networks with renewable distributed generators: A comprehensive survey," *Electric Power Systems Research*, 2009, vol 79, pp. 984–992.
20. A Baggini, *Handbook of Power Quality*, John Wiley & Sons Ltd, UK(2008), pp. 545–546.
21. J Manson, R Targosz, "European Power Quality Survey Report," 2008, pp. 3–15.
22. L Yufeng, "Evaluation of dip and interruption costs for a distribution system with distributed generations," ICHQP2008.
23. N.G. Hingorani, "Introducing custom power," IEEE Spectrum, 1995, vol. 32(6), pp. 41–48.
24. A Ghosh and G Ledwich, Power quality enhancement using custom power devices, Kluwer Academic, 2002.
25. A Ghosh, "Compensation of Distribution System Voltage Using DVR," *IEEE Trans on power delivery*, 2002, vol. 17(4), pp. 1030–1036.
26. H Fujita, H Akagi, "The Unified Power Quality Conditioner: The Integration of Series- and Shunt-Active Filters," *IEEE Trns on power electronics*, 1998, vol. 13, no. 2, pp. 315–322.

27. A Arulampalam, M. Barnes, "Power quality and stability improvement of a wind farm using STATCOM supported with hybrid battery energy storage." Generation, Transmission and Distribution, *IEE Proceedings*, 2006, vol. 153(6): 701–710.
28. Z. Chen, F. Blaabjerg, Y. Hu, "Voltage recovery of dynamic slip control wind turbines with a STATCOM," IPEC05, vol. S29(5), pp. 1093–1100.
29. S.M. Muyeen, R Takahashi, T Murata, J Tamura, M H Ali, "Application of STATCOM/BESS for wind power smoothening and hydrogen generation," *Electric Power Systems Research*, 2009, vol. 79, pp. 365–373.
30. Chung, Y. H., H. J. Kim, "Power quality control center for the microgri system," PECon 2008.
31. M Hosseinpour, Y Mohamadrezapour, S Torabzade, "Combined operation of Unifier Power Quality Conditioner and Photovoltaic Array," *Journal of Applied Sciences*, 2009, v-9(4), pp. 680–688.
32. Jayanti, N. G., M. Basu, "Rating requirements of the unified power quality conditioner to integrate the fixed speed induction generator-type wind generation to the grid." *Renewable Power Generation*, IET, 2009, vol. 3(2): 133–143.
33. A. Domijan, A. Montenegro, "Simulation study of the world's first distributed premium power quality park," *IEEE Trans on Power Delivery*, 2005, vol. 20, pp. 1483–1492.

11

Voltage Transients and Insulation Coordination in Distributed Photovoltaic Grid Power Transformers

Voltage transients are created in distributed photovoltaic grid transformers (DPV-GTs) due to changing weather conditions. The amount of irradiance that impinges on the PV panels is inversely proportional to the cloud coverage. Other situations that can affect the voltage transients are related to the kinds of load attached to such DPV grids. The existence of nonlinear loads can cause proportional transients that subject the insulation of the transformers to unusual stresses that the designer has to be cognizant of to improve the operation of such transformers.

11.1 Insulation Coordination

In its simplest form, insulation coordination is the selection of insulation strength. In a few more words, insulation coordination is a series of steps used to select the dielectric strength of equipment in relation to the operating voltages and transient over-voltages that can appear on the system for which the equipment is intended. Many factors are taken into account during the selection process, including the service environment, insulation withstand characteristics, arrester characteristics, and in some cases, the probability of potential surges.

Self-restoring insulation: Insulation that, after a short time, completely recovers its insulating properties after a disruptive discharge during test

Non-self-restoring insulation: Loses its insulating properties, or does not recover them completely, after a disruptive discharge during test

Flashover rate: The rate at which an insulator flashes over on a system from lightning or switching. For a line study, this rate along with the back flashover rate (BFR) determine the outage rate of a line.

Shielding failure rate (SFR): The shielding failure rate is the number of strikes that terminate on the phase conductors. If the voltage produced by a strike to the phase conductors exceeds the line CFO (critical flashover voltage), flashover occurs.

Back flashover rate (BFR): The BFR is the number of lightning strikes that terminate on towers or shield wires and result in insulator flashover. The current impulse raises the tower voltage, and in turn this generates a voltage across the line insulation. If the voltage across the line insulators exceeds the insulation strength, a back flashover can be expected from the tower onto the phase conductor.

11.2 Data Required for Insulation Coordination Study

The data collected for a study are dependent on the purpose of the study, but the following are examples of required data:

1. Collect BIL, CFO data of all insulation.
2. Collect arrester characteristics and installed locations if applicable.
3. Obtain one line diagram of the system with distances between all insulators and arresters.
4. Obtain insulator counts and locations in particular if a switching study is the goal.
5. Note the region of the country.
6. Obtain lightning data for the area of analysis.
7. Note ground resistances where possible.

11.3 Insulation Coordination Standards

The IEEE standards that describe insulation coordination methods and definitions are C62.82.1 and C62.82.2. Prior to 2010 these standards were referred to as IEEE 1313.1 and 1313.2.

The IEC standards that describe insulation coordination methods and definitions are IEC60071-1, 60071-2, and 60071-4.

As described in Chapter 15, the insulation coordination schemes can be employed to mitigate large voltage fluctuations caused by natural causes like lightning strikes.

Impulse voltage distribution is drastically affected due to high ground capacitances and low series capacitance:

$$\alpha = \sqrt{(C_g/C_s)} \tag{11.1}$$

where C_g is the ground capacitance and C_s is the series capacitance. For disc windings or the helical windings, the value of α is generally about 2 to 2.5. This gives rise to nonuniform voltage distribution along the length of the winding. A designer will always strive to lower the α value to near 1.0 to make the voltage distribution as uniform as possible along the length of the winding. This helps to have a uniform paper covering on the bare conductor of the winding. Most of these problems are minimized by intelligent design of windings. One such way to make the voltage distribution along the length of the winding uniform is by using low series capacitor windings. In some voltage ratings as high as 66 kV, the interleaved winding structure is justifiable.

11.4 Voltage Flicker Concerns

Any dramatic change in current will cause a dramatic change in voltage. This can occur whenever the generator main breaker opens, or during clouding. Flicker is limited to 2 V (2.5%) on a 120 V base in urban areas or 5 V (4.17%) on a 120 V base in rural areas.

The transients are of two types:

1. Voltage dips: a sudden reduction in voltage at a particular point in a circuit followed by a voltage recovery after a short period of time in a few seconds to a few cycles of the system voltage frequency. This could happen due to switching of loads. This voltage dip is generally between 90% and 1% of the nominal system voltage. These values are defined in IEEE P1433 as 0.9 pu to 0.1 pu of the RMS value of the nominal voltage of the system for a duration of 0.5 cycle to 1 minute. Anything less than 90% of a voltage is not declared as a voltage dip. This can last between 10 msec to several seconds up to 1 minute. A voltage dip is caused due to an increase in circuit currents as in the case of short circuits or switching of loads. The former is generally associated with the starting of induction motors where the starting current is six to seven times that of the full-load current.

2. Voltage swell: a temporary increase of voltage at a point in the network above a certain predefined threshold, typically set at 1.1 pu. These are also defined by their magnitude and duration. Voltage swells are generally associated with system faults as in the case of an increase in voltage levels in the unfaulted phases when a single line to ground fault occurs. At the same time these can be caused by the switching of large loads as in the case of the switching of large capacitor banks. The levels of the voltage swells are dependent on the location of the faults, system impedance, and grounding. Most

of the solar system networks in the United States are grounded, and thus the intensity and frequency of occurrence in the US systems is less as compared to the European systems where generally the solar systems may not be grounded.

11.5 Effect of Voltage Variation on Power Flow in Grid-Tied DPV-GT Systems

Generally with grid-tied DPV-GTs, the voltage variation due to high PV penetration (close to 90%) is not an issue, but with microgrids or weak grids, the effect may be pronounced. A new approach has been studied [2] to evaluate the power in such fluctuating irradiance conditions by using spectral analysis of the irradiance time series. Wavelet analysis helps to segregate the fractal distribution of cloud cover physically distributed in a given topology. The node voltage variation is found to be within specified limits of local standards, yet additional limits need to be added due to distributed generation if encountered in the system.

11.6 Voltage Variation Mitigation

Voltage fluctuations due to PV penetration can cause a 20% increase in voltage magnitude, and with a volt-VAR control this fluctuation can be flattened as shown by recent studies conducted by EPRI [1].

Mitigation of voltage fluctuation can be achieved with storage devices that may be an option especially in cable grids with a high series resistance. In mainly inductive grid circuits with interconnecting DPV-GTs, this can be achieved by injection of reactive power by the inverter circuit. Total voltage fluctuations can reach up to 10% above normal voltage. The additional 5% mitigation of the voltage can also be achieved through the use of supercapacitors, with the price of such capacitors being about 20% of the DPV-GT system price.

Bibliography

1. Distribution systems and the integration of solar PV, Jeff Smith, Senior Project Manager, EPRI, Tennessee Valley Solar Solutions Conference, April 10, 2012.

2. Woyte, Achim, Thong, Vu Van, Belmans, Ronnie, Nijs, Johan, Voltage fluctuations on distribution level introduced by photovoltaic systems, IEEE.

3. Murata, A., Otani, K., An analysis of time-dependent spatial distribution of output power from very many PV power systems installed on a nation-wide scale in Japan, *Solar Energy Materials and Solar Cells* 47, 1997, 197–202.

4. Wiemken, E., Beyer, H. G., Heydenreich, W., Kiefer, K., Power characteristics of PV ensembles: Experiences from the combined power production of 100 grid connected PV systems distributed over the area of Germany, *Solar Energy* 70, 2001, 513–518.

5. Jewell, W., Ramakumar, R., The effects of moving clouds on electric utilities with dispersed photovoltaic generation, *IEEE Transactions on Energy Conversion* EC-2, 570–576, December 1987.

6. Jewell, W. T., Ramakumar, R., Hill, S. R., A study of dispersed photovoltaic generation on the PSO system, *IEEE Transactions on Energy Conversion* 3(3), September 1988, 473–478.

7. Jewell, W. T., Unruh, T. D., Limits on cloud-induced fluctuation in photovoltaic generation, *IEEE Transactions on Energy Conversion* 5(1), March 1990, 8–14.

8. Kern, E. C., Russel, M. C., Spatial and temporal irradiance variations over large array fields, *Proceedings of the 20th IEEE Photovoltaic Specialists Conference*, 1988, pp. 1043–1050.

9. Willis, H., Ed., *Distributed power generation: Planning and evaluation*, CRC Press, Boca Raton, FL, 2000.

10. Chalmers, S., Hitt, M., Underhill, J., Anderson, P., Vogt, P., Ingersoll, R. The effect of photovoltaic power generation on utility operation, *IEEE Transactions on Power Apparatus and Systems*, PAS-104, 1985, pp. 524–530.

11. Patapoff, N., Mattijetz, D. Utility interconnection experience with an operating central station MW-sized photovoltaic plant, *IEEE Transactions on Power Systems and Apparatus*, PAS-104, 1985, pp. 2020–2024.

12. Jewell, W., Ramakumar, R., Hill, S. A study of dispersed PV generation on the PSO system, *IEEE Transactions on Energy Conversion* 3, 1988, 473–478.

13. Cyganski, D., Orr, J., Chakravorti, A., Emanuel, A., Gulachenski, E., Root, C., Bellemare, R. Current and voltage harmonic measurements at the Gardner photovoltaic project, *IEEE Transactions on Power Delivery* 4, 1989, 800–809.

14. EPRI report EL-6754, Photovoltaic generation effects on distribution feeders, Volume 1: Description of the Gardner, Massachusetts, twenty-first century PV community and research program, 1990.

15. Garrett, D., Jeter, S. A photovoltaic voltage regulation impact investigation technique: Part I—Model development, *IEEE Transactions on Energy Conversion* 4, 1989, 47–53.

16. Baker, P., McGranaghan, M., Ortmeyer, T., Crudele, D., Key, T., Smith, J. *Advanced grid planning and operation*. NREL/SR-581-42294. Golden, CO: National Renewable Energy Laboratory, January 2008.

17. Jewell, W., Unruh, T. Limits on cloud-induced fluctuation in photovoltaic generation, *IEEE Transactions on Energy Conversion* 5(1), March 1990, 8–14.

18. Imece et al., tests on SunSine inverter.

19. Asano, H., Yajima, K., Kaya, Y. Influence of photovoltaic power generation on required capacity for load frequency control, *IEEE Transactions on Energy Conversion* 11, 1996, 188–193.

20. Povlsen, A., International Energy Agency Report IEA PVPS T5-10: 2002, February 2002 (www.iea.org).

21. Kroposki, B., Vaughn, A., DG power quality, protection, and reliability case studies report, NREL/SR-560-34635. Golden, CO: National Renewable Energy Laboratory. General Electric Corporate R&D, 2003.
22. Miller, N., Ye, Z., Report on distributed generation penetration study. NREL/SR-560-34715. Golden, CO: National Renewable Energy Laboratory, 2003.
23. Kersting, W., *Distribution system modeling and analysis*, CRC Press, Boca Raton, FL, 2002.
24. Union for the Coordination of Transmission of Electricity, Final Report of the Investigation Committee on the 28 September 2003 Blackout in Italy, 2004, p. 121.
25. Quezada, V., Abbad, J., San Román, T., Assessment of energy distribution losses for increasing penetration of distributed generation, *IEEE Transactions on Power Systems* 21(2), May 2006, 533–540.
26. DISPOWER: Distributed generation with high penetration of renewable energy sources, Final Public Report, 2006 (www.pvupscale.org).
27. Thomson, M., Infield, D., Impact of widespread photovoltaics generation on distribution systems, *IET Journal of Renewable Power Generation* 1, 2007, 33–40.
28. Ueda, Y., et al., Performance ratio and yield analysis of grid-connected clustered PV systems in Japan, *Proceedings of the Fourth World Conference on Photovoltaic Energy Conversion*, pp. 2296–2299.
29. Luque, A., Hegedus, S., *Handbook of photovoltaic science and engineering*, Wiley, New York, 2003.
30. Short, T. A., *Electric power distribution handbook*, CRC Press, Boca Raton, FL, 2004.
31. Watson, N., Arrillaga, J., Power systems electromagnetic transients simulation, Institution of Electrical Engineers, 2002.
32. ANSI/IEEE Standard 1547-2003: IEEE standard for interconnecting distributed resources with electric power systems.
33. Ronan, E., Sudhoff, S., Glover, S., Galloway, D., A power electronic-based distribution transformer, *IEEE Transactions on Power Delivery* 17, 2002, 537–543.
34. Hatta, H., Kobayashi, H., A study of centralized voltage control method for distribution system with distributed generation, *Proceedings of the 19th International Conference on Electricity Distribution* (CIRED), May 2007, paper 0330, 4 pages.
35. Okada, N., Verification of control method for a loop distribution system using loop power flow controller, *Proceedings of the 2006 IEEE Power Systems Conference and Exposition*, October 29–November 1, 2006, pp. 2116–2123.

Chapter 11 Problems

1. A 1/2 MVA 22 kV (±10% variation)/433 V, Y/Y oil-filled three-phase transformer has the following details:

   LV winding:

   a. ID: 207 mm

   b. Mean diameter: 222 mm

   c. OD: 244 mm

      d. Axial length: 700 mm

      e. Volts per turn: 12.61

Radial clearance between LV and HV: 22 mm

HV winding:

      a. ID: 288 mm

      b. Mean diameter: 338 mm

      c. OD: 390 mm

      d. Axial length: 700 mm (with 109 discs having eight turns per disc)

      e. Volts per turn: 12.61

  A. Draw the mmf diagram for the transformer windings at normal tap (i.e., at 100% rated voltage on HV winding) for a guaranteed impedance of 7.5%.

  B. Calculate the % reactance of the transformer at the normal tap (i.e., at 100% rated voltage on the HV winding).

2. The transformer in Problem 1 has to be designed with interleaved winding in the HV as compared to an ordinary continuous double disc.

    a. Draw the winding diagrams for the HV disc sections with interleaving as well as double disc configuration.

    b. Draw the distribution of the voltage on the HV winding with interleaving and CDD configurations (i.e., voltage versus length of winding for both α).

    c. What is the improvement in α by the use of the interleaving configuration? Give the answer with a numerical value.

12

Inverter Circuit Coordination with a Distributed Photovoltaic Grid Power Transformer

Inverter technology has been slow to advance as related to the capacity of handling power, because it is an electronic technology. The maximum power rating is at the moment 500 KVA. For higher capacity requirements single units of 500 KVA are assembled in parallel. It remains to be seen whether this comparative disadvantage will be a fatal flaw in the advancement of solar technology to the same level as wind farms in the renewable energy arena. Presently higher rating units up to 1000 KVA are being manufactured to suit the penetration of photovoltaic (PV) systems.

The duty cycle seen in solar farms may not be as severe as seen in wind farms, but solar power has its share of special considerations that affect transformer design. Those engaged in harnessing solar energy need to pay heed to these special needs to ensure that solar installation is cost effective and reliable. To fully understand the effects of inverter technology on the design and operation of distributed photovoltaic grid transformers (DPV-GTs), one has to understand the working of the subassemblies shown in Figure 12.1.

12.1 Inverter Definition

The energy, or power, provided by utilities to industry, residences, and businesses comes from the *infinite electrical grid* and is commonly referred to as alternating current (AC). Some renewable energy power generation systems, such as solar, produce direct current (DC).

Inverters convert DC current into AC current, making it suitable for this alternate energy source harnessed from the sun to power our homes and businesses. Over the past century, inverters have evolved from basic electrical concepts to complex combinations of power electronics and digital controls.

12.2 Inverter History

Inverters began being used in the late nineteenth century as electromechanical devices in the form of rotary converters or motor-generator sets (MG-set).

FIGURE 12.1
(See color insert) Sun-Solar panel-inverter DPV-GT feeder lines.

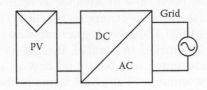

FIGURE 12.2
DC to AC conversion from a DC solar source.

It is easy to understand the concept of an inverter. Thus, if the connections to a MG-set (converter) are inverted and if DC is inserted into the circuit, the inverter circuit produces AC at the output as shown in Figure 12.2.

Hence, an inverter is an inverted converter. In the early twentieth century, vacuum tubes and gas-filled tubes were used to perform this conversion. In 1947, transistors with a low voltage threshold were used to achieve this conversion, but it restricted the transistor's use in power conversion due to the low power handling capacity. In 1957 with the advent of the silicon controlled rectifier (SCR) or thyristor, power conversion could be achieved at moderate power handling capacities. Later, additional research in power electronics initiated the transition to high-power solid-state inverter circuits. Of particular interest for this application was the gate turn-off thyristor (GTOT). Further research led to Insulated-gate bipolar transistor (IGBT) and integrated-gate commutated/control thyristor (IGCT) technologies to be discussed later. These power electonic devices are presently suitable to handle power capacities up to 1000 KVA. Many leading manufacturers in the United States and Europe have started to provide practical configurations of inverters using IGBTs and IGCTs.

12.3 Inverter Technology

Many manufacturers have been using inverter technology for over thirty years to control the speed of AC motors via variable speed drives (VSDs).

Solar inverters can be viewed as essentially one half of industrial drives, working backwards. Solar inverters use the same proven IGBT/IGCT design as industrial drives.

12.3.1 Variable Speed Drive versus Solar Inverter

A typical PV solar system consists of a DC input from solar cells to MV transformer that can produce a three-phase AC output, using silicon technology components like IGBTs and/or IGCTs, while a variable speed drive utilizes a similar set of silicon technology components to utilize circuitry to control the speed of motors, either DC (series, shunt, or compound machines) or AC machines like induction motors.

12.3.2 Solar Inverters—Grid Tied versus Non-Grid Tied

Grid tied: These solar inverters are connected directly to the utility power grid. Generally, there are no backup power storage systems like batteries. Although a recent increase in solar energy capacities added worldwide to supplement the battery storage systems are slowly being added in the mix. Such solar inverters also require grid-related monitoring, feedback, and safety features which are essential features when used for selling power to the utility.

Non-grid tied: These solar inverters operate independent of the utility power grid and in general have a backup system (i.e., batteries, diesel generator, etc.). These do not require extensive monitoring, but some local monitoring is favored for safety reasons. Such a system is normally installed when utility grid power is not easily accessible, mainly during a power outage, or cannot be installed in a cost-effective manner in remote locations.

12.3.3 Solar Inverter Features and Characteristics (Grid-Tied)

Solar inverters differ from standard inverters used in variable speed drives only in their auxiliary controls and monitoring capabilities (usually software).

In standard IGBT technology, in addition to the normal electrical codes (NEC, NEMA, etc.), the features of the solar inverter are driven, for the most part, by Underwriters Laboratory (UL) and the California Energy Commission (CEC). The guidelines include

1. Grid-tied inverters must include control and monitoring features (as listed here).
2. Grid-tied inverters must have s maximum power point tracking (MPPT) system.

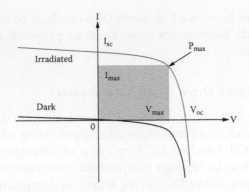

FIGURE 12.3
Typical *i-v* characteristics of an MPPT system.

3. Power monitoring of the following parameters is required: control of voltage, current, power factor, frequency, harmonics, and synchronization with the utility grid.

4. Anti-islanding (grid-tied systems) is necessary as described in Chapters 6 and 17.

5. Ground-fault monitoring is essential.

6. Enclosures should be suitable for a given environment as described in related IEEE codes and as per the other NEMA IP standards.

7. Communication ports to isolate and avoid islanding events have to be provided to prevent failure of the system and to prevent lives from being endangered.

12.3.4 Solar Inverters—Maximum Power Point Tracking

Generally on the DC side, the power output of the solar modules varies as a function of the voltage in such a way that power generation can be optimized by varying the system voltage to find the maximum power point. Most solar inverters therefore incorporate MPPT. Figure 12.3 shows the *i-v* characteristics of this mechanism with P_{max} as the maximum power on the *i-v* curve for the irradiated and dark conditions, respectively.

$$P_{max} = V_{max} \times I_{max} \tag{12.1}$$

12.3.5 Solar Inverters—Power Monitoring

Power monitoring of solar inverters should satisfy the following conditions:

1. Good power grid citizens should preserve energy when not needed.

2. Monitoring must comply with utility electrical standards (efficiency and losses).

3. Standards followed are usually UL1741 or CEC, but they can be determined by the local utility.

4. The output power needs to be clean, undistorted, and in phase or synchronized with the utility grid. It should be a pure sine wave.

5. Most solar inverters have a monitoring system that will sense the utility grid waveform (voltage, current, power factor, frequency) and provide an output to correspond. Care has to be taken to maintain the sinusoid nature, and the power quality should be maintained.

6. Typical modern grid-tied inverters have a fixed unity power factor, which means the output voltage and current are perfectly lined up, and its phase angle is within 1 degree of the AC power grid.

12.4 DC Bias Caused by Inverters

Inverter circuits typically will introduce some amount of DC offset voltage also called DC bias. This can lead to saturation of the magnetic flux in the core of the transformer. In order to avoid this phenomenon different kinds of inverter circuits are utilized:

1. Six-pole inverter circuit
2. Twelve-pole inverter circuit
3. Twenty-four pole inverter circuits
4. Pulse width modulated (PWM) inverter circuits using IGBTs

Each type of circuit used as above will also lead to some commutation. Commutation is a phenomenon when more than one circuit component carries current simultaneously causing a short circuit for a fraction of the duty cycle. This may prove dangerous if allowed to be sustained for a longer period (>50%) of the duty cycle.

12.5 Typical DPV Generation Systems and Their Specifics in Relation to the Transformers

The common components of a grid-tied DPV System are the solar generator, inverter, electric RLC filter, and switch that connects to the grid as shown in Figure 12.4.

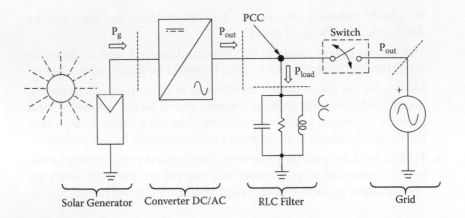

FIGURE 12.4
Typical DPV generation system single line diagram.

The AC module depicted in Figure 12.4 is the integration of the inverter and PV module into one electrical device. It removes the mismatch losses between PV modules since there is only one PV module, as well as supports optimal adjustment between the PV module and the inverter and, hence, the individual MPPT. It includes the possibility of an easy enlarging of the system, due to the modular structure. The opportunity to become a "plug-and-play" device, which can be used by persons without any knowledge of electrical installations, is also an inherent feature. On the other hand, the necessary high-voltage amplification may reduce the overall efficiency and increase the price per watt because of more complex circuit topologies. Presently, the AC module is intended to be mass produced, which leads to low manufacturing cost and low retail prices. The present solutions use self-commutated DC–AC inverters, by means of IGBTs or MOSFETs.

12.6 Types of Converter Topologies

There are three types of converters: central inverters, string inverters, and module-oriented or module-integrated inverters.

The string and multistring systems are the combination of one or several PV strings with a ground as shown in Figure 12.4. The inverters should be of the single- or dual-stage type with or without an embedded HF transformer. The input voltage may be high enough to avoid voltage amplification. This requires roughly 16 PV modules in series for European systems.

The total open-circuit voltage for 16 PV modules may reach as much as 720 V, which calls for a 1000 V metal oxide semiconductor field effect transistor (MOSFET)/insulated-gate bipolar transistor (IGBT)/integrated gate

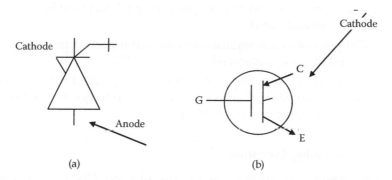

FIGURE 12.5
(a) Insulated-gate bipolar transistor (IGBT); (b) integrated gate commutated/control thyristor (IGCT).

commutated/control thyristor (IGCT) in order to allow for a 75% voltage de-rating of the semiconductors. The normal operation voltage is, however, as low as 450 to 510 V. The possibility of using fewer PV modules in series also exists if a DC–DC converter or line-frequency transformer is used for voltage amplification.

12.7 Inverter Technology

Some silicon devices are shown in Figure 12.5. IGBT is a conventional device used over the past three decades. IGCT is a newer device and can be used for energy conversion methodologies efficiently.

12.8 Solar Inverters—Anti-Islanding

12.8.1 Anti-Islanding (Grid-Tied Systems)

Most inverters detect the islanding condition by looking for some combination of the following:

1. A sudden change in system frequency
2. A sudden change in voltage magnitude
3. A sudden change in the df/dt (rate of change of frequency)
4. A sudden increase in active output power (kW) well beyond the expected "normal" level

5. A sudden change in reactive output power (kVAR) well beyond an expected "normal" level

6. UL 1741 prescribes the requirements for anti-islanding protection in the United States (harmonized with IEEE 1547).

7. Simulates the utility using a phase-locked loop (PLL) with a small amount of positive feedback in its control loop to quickly and continuously check the grid connectivity

12.8.2 Anti-Islanding Exception

New requirements for utility-scale PV installations ("behind-the-fence") include 1000 VDC instead of 600 VDC maximum, no UL, and no anti-islanding is required since the control is at the substation level. In reality, the "opposite" of anti-islanding exists. Also applied are low voltage and fault ride-through of the Federal Energy Regulatory Commission (FERC) and low-voltage ride-through order 661A, and Western Electricity Coordinating Council (WECC) low-voltage ride-through criterion PRC-024-1-CR Power Factor Correction (VAR control).

12.9 Grid-Tie Inverter or Synchronous Inverters

A *grid-tie inverter* (GTI) or *synchronous inverter* is a special type of power inverter (see Figures 12.6 and 12.7) that converts direct current (DC) electricity into alternating current (AC) and feeds it into an existing electrical grid. GTIs are often used to convert direct current produced by many renewable energy sources, such as solar panels or small wind turbines, into the alternating current used to power homes and businesses. The technical name for a grid-tie inverter is *grid-interactive inverter*. Grid-interactive inverters typically cannot be used in stand-alone applications where utility power is not available. During a period of overproduction from the generating source, power is routed into the power grid, thereby being sold to the local power company. During insufficient power production, it allows for power to be purchased from the power company. Residences and businesses that have a grid-tied electrical system are permitted in many countries to sell their energy to the utility grid. Electricity delivered to the grid can be compensated in several ways. *Net metering* is where the entity that owns the renewable energy power source receives compensation from the utility for its net outflow of power. So for example, if during a given month a power system feeds 500 kilowatt-hours into the grid and uses 100 kilowatt-hours from the grid, it would receive compensation for 400 kilowatt-hours. In the United States, net metering policies vary by jurisdiction. Another policy is a feed-in

FIGURE 12.6
(See color insert) Inverter for grid-connected PV.

tariff, where the producer is paid for every kilowatt hour delivered to the grid by a special tariff based on a contract with a distribution company or other power authority.

In the United States, grid-interactive power systems are covered by specific provisions in the National Electric Code, which also mandates certain requirements for grid-interactive inverters.

12.9.1 Typical Operation

Inverters take DC power and invert it to AC power so it can be fed into the electric utility company grid. The GTI must synchronize its frequency with that of the grid (e.g., 50 or 60 Hz) using a local oscillator and limit the voltage to no higher than the grid voltage. A high-quality modern GTI has a fixed unity power factor, which means its output voltage and current are perfectly lined up, and its phase angle is within 1 degree of the AC power

FIGURE 12.7
(See color insert) Example of large, three-phase inverter for commercial and utility-scale grid-tied PV systems.

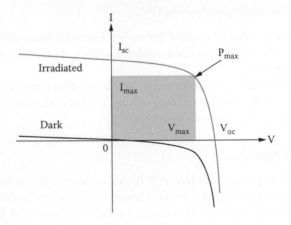

FIGURE 12.8
(See color insert) Current-voltage characteristics of a solar cell: the area of the rectangle gives the output power; and P_{max} denotes the maximum power point.

grid. The inverter has an on-board computer that will sense the current AC grid waveform and will output a voltage to correspond with the grid.

Grid-tie inverters are also designed to quickly disconnect from the grid if the utility grid goes down. This is a NEC requirement that ensures that in the event of a blackout, the GTI will shut down to prevent the energy it produces from harming any line workers who are sent to fix the power grid.

Properly configured, a GTI enables a home owner to use an alternative power generation system like solar or wind power without extensive rewiring and without batteries. If the alternative power being produced is insufficient, the deficit will be sourced from the electricity grid.

Restart timing for each inverter is to be spaced one minute apart to ease possible power quality impacts. A typical v-i characteristic of a solar cell is shown in Figure 12.8.

12.9.2 Technology

Grid-tie inverters that are available on the market today use a number of different technologies. The inverters may use the newer high-frequency transformers, conventional low-frequency transformers, or go without a transformer. Instead of converting direct current directly to 120 or 240 volts AC, high-frequency transformers employ a computerized multistep process that involves converting the power to high-frequency AC and then back to DC and then to the final AC output voltage. Transformerless inverters, lighter and more efficient than their counterparts with transformers, are popular in Europe. However, transformerless inverters have been slow to enter the US market because historically there have been concerns about having transformerless electrical systems feeding into the public utility grid. Because of the lack of galvanic isolation between the DC and AC circuits, dangerous DC faults could be transmitted to the AC side. However, since 2005, the NFPA's NEC allows transformerless (or nongalvanic) inverters by removing the requirement that all solar electric systems be negatively grounded and by specifying new safety requirements. The VDE 0126-1-1 and IEC 6210 also have been amended to allow and define the safety mechanisms needed for such systems. Primarily, residual or ground current detection is used to detect possible fault conditions. Isolation tests are performed to ensure DC to AC separation.

Most GTIs on the market include a maximum power point tracker (MPPT) on the input side that enables the inverter to extract a maximum amount of power from its intended power source. Since MPPT algorithms differ for solar panels and wind turbines, specially made inverters for each of these power sources are available.

12.9.3 Characteristics

Inverter manufacturers publish datasheets for the inverters in their product lines. While the terminology and content will vary by manufacturer, data sheets generally include the following information:

- *Rated output power*: This value will be provided in watts or kilowatts. For some inverters, an output rating may be provided for different output voltages. For instance, if the inverter can be configured for either 240 VAC or 208 VAC output, the rated power output may be different for each of those configurations.

- *Output voltage(s)*: This value indicates to which utility voltages the inverter can connect. For smaller inverters that are designed for residential use, the output voltage is usually 240 VAC. Inverters that target commercial applications are rated for 208, 240, 277, 400, or 480 VAC and may also produce three-phase power.

- *Peak efficiency*: The peak efficiency represents the highest efficiency that the inverter can achieve. Most GTIs on the market as of July 2009 have peak efficiencies of over 94%, some as high as 96%. The energy lost during inversion is for the most part converted into heat. This means that in order for an inverter to put out the rated amount of power, it will need to have a power input that exceeds the output. For example, a 5000 W inverter operating at full power at 95% efficiency will require an input of 5263 W (rated power divided by efficiency). Inverters that are capable of producing power at different AC voltages may have different efficiencies associated with each voltage.

- *CEC weighted efficiency*: This efficiency is published by the California Energy Commission (CEC) on its GoSolar website. In contrast to peak efficiency, this value is an average efficiency and is a better representation of the inverter's operating profile. Inverters that are capable of producing power at different AC voltages may have different efficiencies associated with each voltage.

- *Maximum input current*: This is the maximum amount of direct current that the inverter will use. If a DC power source, such as a solar array, produces an amount of current that exceeds the maximum input current, that current will not be used by the inverter.

- *Maximum output current*: The maximum output current is the maximum continuous alternating current that the inverter will supply. This value is typically used to determine the minimum current rating of the over-current protection devices (e.g., breakers and fuses) and disconnects required for the output circuit. Inverters that are capable of producing power at different AC voltages will have different maximum outputs for each voltage.

- *Peak power tracking voltage*: This represents the DC voltage range in which the inverter's maximum point power tracker will operate. The system designer must configure the strings optimally so that during the majority of the year, the voltage of the strings will be within this range. This can be a difficult task since voltage will fluctuate with changes in temperature.

- *Start voltage*: This value is not listed on all inverter datasheets. The value indicates the minimum DC voltage that is required in order for the inverter to turn on and begin operation. This is especially important for solar applications, because the system designer must be sure that there are a sufficient number of solar modules wired in series in each string to produce this voltage. If this value is not provided by the manufacturer, system designers typically use the lower band of the peak power tracking voltage range as the inverter's minimum voltage.
- *Ingress protection (IP) rating*: The ingress protection rating or IP Code classifies and rates the level of protection provided against the ingress of solid foreign objects (first digit) or water (second digit)—a higher digit means greater protection. In the United States, the NEMA rating is used similarly to the international rating. Most inverters are rated for outdoor installation with IP45 (no dust protection) or IP65 (dust tight), or in the United States, NEMA 3R (no windblown dust protection) or NEMA 4X (windblown dust, direct water splash, and additional corrosion protection).

12.10 Solar Micro-Inverter

A *solar micro-inverter*, or *microinverter* or *micro inverter*, converts direct current (DC) electricity from a single solar panel to alternating current (AC) (Figure 12.9). The electric power from several micro-inverters is combined and fed into an existing electrical grid. Micro-inverters contrast with conventional string or central inverter devices, which are connected to multiple solar panels.

FIGURE 12.9
(See color insert) A solar micro-inverter.

Micro-inverters have several advantages over conventional central inverters. The main advantage is that even small amounts of shading, debris, or snow lines in any one solar panel, or a panel failure, do not disproportionately reduce the output of an entire array. Each micro-inverter obtains optimum power by performing maximum power point tracking for its connected panel.

Their primary disadvantages are that they have a higher equipment initial cost per peak watt than the equivalent power in a central inverter, and they are normally located near the panel, where they may be harder to maintain. These issues are, however, balanced by micro-inverters having good durability and simplicity of initial installation.

Micro-inverters that accept DC input from two solar panels, rather than one, are a recent development. They perform independent maximum power point tracking on each connected panel. This reduces the equipment cost and makes photovoltaic (PV) systems based on micro-inverters comparable in cost with those using string inverters.

12.11 String Inverters

Solar panels produce direct current at a voltage that depends on the module's design and the lighting conditions. Modern panels using 6″ cells normally contain 60 cells and produce a nominal 30 volts. For conversion into AC, panels are connected in series to produce effectively a single, large array with a nominal rating of around 300 to 600 VDC. The power is then run to an inverter, which converts it into standard AC voltage, typically 240 VAC/60 Hz for the North American market, or 220 VAC/50 Hz in Europe.

The main problem with this *string inverter* approach is that the string of panels will act as if it was a single, larger panel rated to the worst of the individual panels within it. For instance, if one panel in a string has 5% higher resistance due to a minor manufacturing defect, the string as a whole will perform 5% worse (or thereabouts). This situation is dynamic; if a panel is shaded, its output drops dramatically, affecting the output of the string as a whole even if the other panels are not shaded. Even slight changes in orientation can cause mismatches in output in this fashion.

Additionally, the efficiency of a panel's output is strongly affected by the load the inverter places on it. In order to maximize production, inverters use a technique known as maximum power point tracking (MPPT) to ensure optimal collection by adjusting the applied load. However, the same issues that cause output to vary from panel to panel affect the proper load that the MPPT system should apply. If a single panel is operating at a different point, a string inverter can only see the overall change and will move the MPPT point to match. This will result in not just the losses from the shadowed panel, but all of the other panels as well. Shading of as little as 9% of the entire surface array of a PV system can, in some circumstances, lead to a system-wide power loss of as much as 54%.

A further problem, although minor, is that string inverters come in a limited selection of power ratings. This means that a given array will normally upsize the inverter to the next-largest model over the rating of the panel array. For instance, a 10-panel array of 2300 W might have to use a 2500 W or even 3000 W inverter, paying for conversion capability it cannot use. This same effect makes it difficult to change array sizes over time, adding power when funds are available. With micro-inverters, different ratings of solar panels can be added to an array even if they do not match the original types.

Other challenges associated with centralized inverters include the space required to locate the device, as well as heat dissipation requirements. Large central inverters are typically actively cooled. Cooling fans make noise, so the location of the inverter relative to offices and occupied areas must be considered.

12.12 Micro-Inverters

Micro-inverters are small inverters rated to handle the output of a single panel. Modern grid-tie panels are normally rated between 220 and 245 W, but rarely do they produce this in practice, so micro-inverters are typically rated between 190 and 220 W. Because it is operated at this lower power point, many design issues inherent to larger designs simply go away; the need for a large transformer is generally eliminated, large electrolytic capacitors can be replaced by more reliable thin-film capacitors, and cooling loads are so reduced that no fans are needed. The mean times between failures (MTBFs) are quoted in the hundreds of years.

More importantly, a micro-inverter attached to a single panel allows it to isolate and tune the output of that panel. A dual micro-inverter does this for two panels. For example, in the same 10-panel array used as an example above, with micro-inverters any panel that is underperforming will have no effect on the panels around it. In that case, the array as a whole will produce as much as 5% more power than it would with a string inverter. When shadowing is factored in, if present, these gains can become considerable, with manufacturers generally claiming 5% better output at a minimum, and up to 25% better in some cases.

Micro-inverters produce grid-matching power directly at the back of the panel. Arrays of panels are connected in parallel to each other, and then to the grid feed. This has the major advantage that a single failing panel or inverter will not take the entire string offline. Combined with the lower power and heat loads, and improved MTBF, it is suggested that overall array reliability of a micro-inverter-based system will be significantly greater than a string-inverter-based one. This assertion is supported by longer warranties, typically fifteen to twenty-five years, compared with five- or ten-year warranties that are more typical for string inverters. Additionally, when faults occur, they are identifiable to a single point, as opposed to an entire string. This not only makes fault isolation easier, but unmasks minor problems that might never become visible otherwise—a single underperforming panel may not affect a short string's output enough to be noticed.

The main disadvantage of the micro-inverter concept has, until recently, been cost. Because each panel has to duplicate much of the complexity of a string inverter, the costs are marginally greater. This offsets any advantage in terms of simplification of the individual components. As of October 2010, a central inverter costs approximately $0.40 per watt, whereas a micro-inverter costs approximately $0.52 per watt. Like string inverters, economic considerations force manufacturers to limit the number of models they produce; most produce a single model that may be over- or under-size when matched with a particular panel. With steadily decreasing prices, the introduction of dual micro-inverters that accept DC input from two solar panels, and the advent of wider model selections to match PV module output more closely, cost is less of an obstacle, so micro-inverters may now spread more widely. In 2011, the introduction of dual micro-inverters that accept DC input from two solar modules rather than one, reduced equipment costs to the extent that PV systems based on this kind of micro-inverter are comparable in cost with those using string inverters.

Micro-inverters have become common where array sizes are small and maximizing performance from every panel is a concern. In these cases the difference in price-per-watt is minimized due to the small number of panels and has little effect on overall system cost. The improvement in energy collection given a fixed-size array can offset this difference in cost. For this reason, micro-inverters have been most successful in the residential market, where the limited space for panels constrains the array size, and shading from nearby trees or other homes is often an issue. Micro-inverter manufacturers list many installations, some as small as a single panel and the majority under fifty panels.

The micro-inverter concept has been in the solar industry since its inception. However, flat costs in manufacturing, like the cost of the transformer or enclosure, scaled favorably with size and meant that larger devices were inherently less expensive in terms of price per watt. Small inverters were available from companies like Exceltech and others, but these were simply small versions of larger designs with poor price performance and were aimed at niche markets.

In 1991 the U.S. company Ascension Technology started work on what was essentially a shrunken version of a traditional inverter, intended to be mounted on a panel to form an *AC panel*. This design was based on the conventional linear regulator, which is not particularly efficient and dissipates considerable heat. In 1994 they sent an example to Sandia Labs for testing.

FIGURE 12.10
(See color insert) Released in 1993, Mastervolt's Sunmaster 130S was the first true micro-inverter.

In 1997, Ascension partnered with U.S. panel company ASE Americas to introduce the 300 W SunSine panel.

Design of what would today be recognized as a *true* micro-inverter traces its history to work done in the late 1980s by Werner Kleinkauf at the Institut für Solare Energieversorgungstechnik (ISET). These designs were based on modern, high-frequency switching power supply technology, which is much more efficient. His work on *module integrated converters* was highly influential, especially in Europe.

In 1993 Mastervolt introduced their first grid-tie inverter, the Sunmaster 130S, based on a collaborative effort between Shell Solar, Ecofys, and ECN (Figure 12.10). The 130S was designed to be mounted directly to the back of the panel, connecting both AC and DC lines using compression fittings. In 2000 the 130S was replaced by the Soladin 120, a micro-inverter in the form of an AC adapter that allows panels to be connected simply by plugging them into any wall socket.

In 1995, OKE-Services designed a new high-frequency version with improved efficiency, which was introduced commercially as the OK4-100 in 1995 by NKF Kabel, and re-branded for U.S. sales as the Trace Microsine (Figure 12.11). A new version, the OK4All, improved efficiency and had wider operating ranges.

In spite of this promising start, by 2003 most of these projects had ended. Ascension Technology was purchased by Applied Power Corporation, a large integrator. APC was in turn purchased by Schott in 2002, and SunSine production was canceled in favor of Schott's existing designs [18]. NKF ended production of the OK4 series in 2003 when a subsidy program ended. Mastervolt has moved on to a line of *mini-inverters* combining the ease-of-use of the 120 in a system designed to support up to 600 W of panels. Released in 2008, the Enphase M175 model was the first commercially successful micro-inverter, and as of September 2011 the company shipped its millionth unit, growing

FIGURE 12.11
(See color insert) Another early micro-inverter, the OK4E-100 (1995) (E for European, 100 for 100 W).

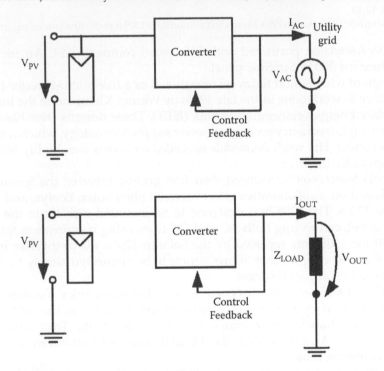

FIGURE 12.12
Converter with feedback tied to grid or load.

from 0% in July 2008 to 30% at the end of July 2011 in the Californian PV residential share market [21]. Inverter with feedback is shown in Figure 12.12.

Since 2009, several companies from Europe to China, including major central inverter manufacturers, have launched micro-inverters, validating the micro-inverter as an established technology and one of the biggest technology shifts in the PV industry in recent years [22].

12.13 Central, Module-Oriented or Module-Integrated, and String Inverters

Central converters connect in parallel and/or in series on the DC side. One converter is used for the entire PV plant (often divided into several units organized in master–slave mode). The nominal power of this topology is up to several megawatts. The module-oriented converters with several modules usually connect in series on the DC side and in parallel on the AC side. The nominal power ratings of such PV power plants are up to several megawatts. In addition, in the module-integrated converter topology, one converter per PV module and a parallel connection on the AC side are used. In this topology, a central measure for main supervision is necessary.

The integration of PV strings (Figure 12.13) of different technologies and orientations is accomplished by following the demands defined by the grid. The EN standard (applied in Europe) allows higher current harmonics, and there are corresponding IEEE and IEC standards.The injection is limited to avoid saturation of the distribution transformers. Limits are rather small (0.5% and 1.0% of rated output current), and such small values can be difficult to measure precisely with the exciting circuits inside the inverters. This can be mitigated with improved measuring circuits or by including a line-frequency transformer between the inverter and the grid.

Some inverters use a transformer embedded in a high-frequency DC–DC converter for galvanic isolation between the PV modules and the grid. This does not, however, solve the problem with DC injection but makes the grounding of the PV modules easier.

- Islanding is the continued operation of the inverter when the grid has been removed on purpose, by accident, or by damage.
- Detection schemes include active and passive.
  - Passive methods monitor grid parameters.
  - Active schemes introduce a disturbance into the grid and monitor the effect.
- In other words, the grid has been removed from the inverter, which then only supplies local loads.
- The NEC 690 standard ensures the system is grounded and monitored for ground faults
- Other electricity boards only demand equipment ground of the PV modules in the case of absent galvanic isolation.
- Equipment ground is the case when frames and other metallic parts are connected to the ground.

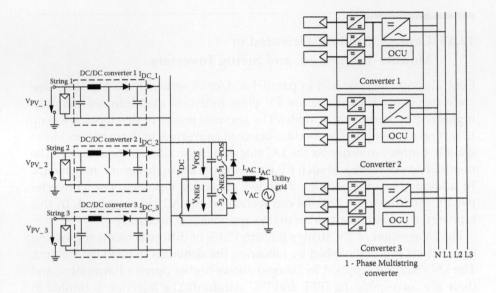

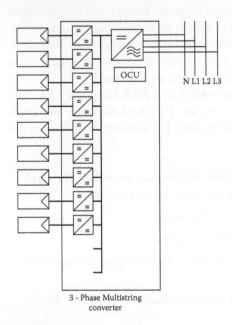

FIGURE 12.13
Multistring inverters.

TABLE 12.1

Harmonic Contents as Specified in Different Standards

Issue	IEC61727 [3]	IEEE1547 [5]	EN6100-3-2 [4]
Nominal power	10 kW	30 kW	16 A × 230 V = 3.7 kW
Harmonic currents	(3–9) 4.0%	(2–10) 4.0%	(3) 2.30 A
(Order - h) Limits	(11–15) 2.0%	(11–16) 2.0%	(5) 1.14 A
	(17–21) 1.5%	(17–22) 1.5%	(7) 0.77 A
	(23–33) 0.6%	(23–34) 0.6%	(9) 0.40 A
		(>35) 0.3%	(11) 0.33%
			(13) 0.21 A
			(15-39) 2.25/h
	Even harmonics in these ranges will be less than 25% of the odd harmonic limits listed.		Approximately 30% of the odd harmonics (see standard).
Maximum current THD	5.0%		—
Power factor at 50% of rated power	0.9	—	
DC current injection	Less than 1.0% of rated output current	Less than 0.5% of rated output current	<0.22 A corresponds to a 50 W half-wave rectifier.
Voltage range for normal operation	85%–110% (196 V–253 V)	88%–110% (97 V–121 V)	—
Frequency range for normal operation	50 + I Hz	59.3 Hz–60.5 Hz	—

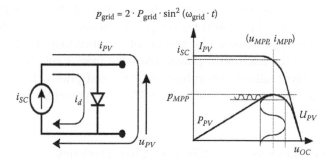

$$p_{grid} = 2 \cdot P_{grid} \cdot \sin^2 (\omega_{grid} \cdot t)$$

FIGURE 12.14
Power in the grid.

12.13.1 Power Injected into Grid

Figure 12.14 shows power in the grid.

- Decoupling is necessary.
- p is instantaneous.
- P is average.

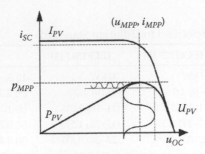

FIGURE 12.15
MPPT for newer PV cells.

12.13.2 Demands Defined by the Photovoltaic Module

The inverters must guarantee that the PV module(s) is operated at the MPP, which is the operating condition where the most energy is captured. This is accomplished with an MPPT. A MPPT is a high-efficiency DC-to-DC converter that presents an optimal electrical load to a solar panel or array and produces a voltage suitable for the load.

PV cells have a single operating point where the values of the current (I) and voltage (V) of the cell result in a maximum power output. These values correspond to a particular resistance, which is equal to V/I as specified by Ohm's law. A PV cell has an exponential relationship between current and voltage, and the maximum power point (MPP) occurs at the knee of the curve, where the resistance is equal to the negative of the differential resistance ($V/I = -dV/dI$). Maximum power point trackers utilize some type of control circuit or logic to search for this point and thus to allow the converter circuit to extract the maximum power available from a cell. New technologies like thin-layer silicon, amorphous silicon, and photo-electro-chemical (PEC) provide maximum power for the PV cells presently manufactured (Figure 12.15).

12.13.3 Maximum Power Point Tracker Characteristics

The following is an example of MPPT characteristics: Ripple voltage should be below 8.5% of the MPP voltage in order to reach a utilization ratio of 98%:

$$\hat{u} = \sqrt{\frac{(k_{PV} - 1) \cdot 2 \cdot P_{MPP}}{3 \cdot \alpha \cdot U_{MPP} + \beta}} = 2 \cdot \sqrt{\frac{(k_{PV} - 1) \cdot P_{MPP}}{\dfrac{d^2 p_{PV}}{du_{PV}^2}}}$$

where u is the amplitude of the voltage ripple, *PMPP* and *UMPP* are the power and voltage at the *MPP*, alpha and beta are the coefficients describing a second-order Taylor approximation of the current, and the utilization ratio *KPV* is given as the average generated power divided by the theoretical MPP power.

12.13.4 High Efficiency

- There is a wide range of input voltage and input power.
- There are very wide ranges as functions of solar irradiation and ambient temperature.

12.13.5 Reliability

Most PV module manufacturers offer a warranty of twenty-five years on 80% of initial efficiency. The main limiting components inside the inverters are the electrolytic capacitors used for power decoupling between the PV module and the single-phase grid. However, the equation assumes a constant temperature, which can be approximated when the inverter is placed indoors and the power loss inside the capacitor is neglected, but certainly not when the inverter is integrated with the PV module, as for the AC module. In the case of a varying temperature, a mean value of (8) must be applied to determine the lifetime.

12.13.6 Topologies of PV Inverters

12.13.6.1 Centralized Inverters

PV modules as series connections (a string) are then connected in parallel, through string diodes (Figure 12.16). There are disadvantages to these inverters.

These have high-voltage DC cables between the PV modules and the inverter, power losses due to a centralized MPPT, mismatch losses between the PV modules, losses in the string diodes, and a nonflexible design where the benefits of mass production could not be reached. The grid-connected stage was usually line commutated by means of thyristors, involving many current harmonics and poor power quality. The large amount of harmonics was the occasion of new inverter topologies and system layouts, in order to cope with the emerging standards that also covered power quality.

12.13.6.2 String Inverters

A reduced version of the centralized inverter single string of PV modules is connected to the inverter, and no losses on string diodes separate MPPTs, increasing the overall efficiency (Figure 12.17). The input voltage may be high enough to avoid voltage amplification. This requires roughly 16 PV modules in series for European systems. The total open-circuit voltage for 16 PV modules may reach as much as 720 V, which calls for a 1000 V MOSFET/IGBT

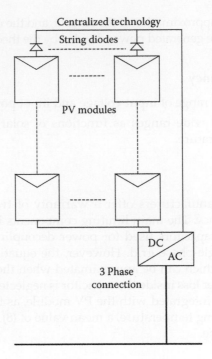

FIGURE 12.16
Centralized inverter configurations.

in order to allow for a 75% voltage de-rating of the semiconductors. The normal operation voltage is, however, as low as 450 to 510 V. The possibility of using fewer PV modules in series also exists, if a DC–DC converter or line-frequency transformer is used for voltage amplification.

12.13.6.3 AC Module

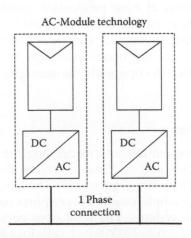

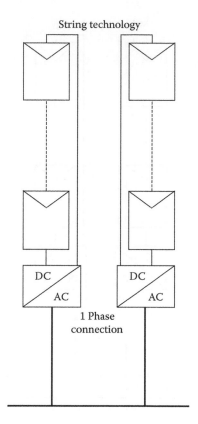

FIGURE 12.17
String converter configuration.

Such an inverter and PV module used as one electrical device leads to no mismatch losses between PV modules. Optimal adjustment of MPPT high-voltage amplification is necessary. The AC module depicted in Figure 12.3 is the integration of the inverter and PV module into one electrical device [7]. It removes the mismatch losses between PV modules since there is only one PV module, as well as supports optimal adjustment between the PV module and the inverter and, hence, the individual MPPT. It includes the possibility of an easy enlarging of the system, due to the modular structure. The opportunity to become a "plug-and-play" device, which can be used by persons without any knowledge of electrical installations, is also an inherent feature. The necessary high-voltage amplification may reduce the overall efficiency and increase the price per watt because of more complex circuit topologies. The AC module is intended to be mass produced, which leads to low manufacturing cost and low retail prices. The present solutions use self-commutated DC–AC inverters, by means of IGBTs or MOSFETs, involving high power quality in compliance with the standards.

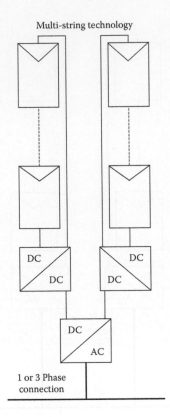

FIGURE 12.18
Multistring inverter configurations.

12.13.7 Future Topologies

12.13.7.1 Multistring Inverters

Multistring inverters are flexible, and every string can be controlled individually (Figure 12.18). The multistring inverter depicted in Figure 12.18 is a further development of the string inverter, where several strings are interfaced with their own DC–DC converter to a common DC–AC inverter [7,28]. This is beneficial, compared with the centralized system, since every string can be controlled individually. Thus, the operator may start his/her own PV power plant with a few modules. Further enlargements are easily achieved since a new string with DC–DC converter can be plugged into the existing platform. A flexible design with high efficiency is hereby achieved.

12.13.7.2 AC Cell Configuration

A general configuration of an AC cell that leads to AC to be connected to the grid is shown in Figure 12.19. One large PV cell is connected to a DC–AC inverter at very low voltage.

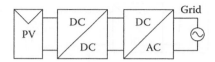

FIGURE 12.19
General configuration of an AC cell that leads to AC to be connected to the grid.

New converter. Concept is really great at very low voltages, 0.5, 1.0 V and 100 W/m², up to an appropriate level for the grid, and at the same time reach a high efficiency. For the same reason, entirely new converter concepts must be designed and manufactured.

12.13.7.3 Classification of Inverter Topologies

The inverter shown in Figure 12.20 is a single-stage inverter that must handle all tasks itself (i.e., MPPT, grid current control, and perhaps, voltage amplification). This is the typical configuration for a centralized inverter, with all the drawbacks associated with it. The inverter must be designed to handle a peak power of twice the nominal power, according to [1]. The DC–DC converter is now performing the MPPT (and perhaps voltage amplification). Dependent on the control of the DC–AC inverter, the output from the DC–DC converters is either a pure DC voltage (and the DC–DC converter is only designed to handle the nominal power), or the output current of the DC–DC converter is modulated to follow a rectified sine wave (the DC–DC converter should now handle a peak power of twice the nominal power). The DC–AC inverter is in the former solution controlling the grid current by means of pulse width modulation (PWM) or bang-bang operation. In the latter, the DC–AC inverter is switching at line frequency, "unfolding" the rectified current to a full-wave sine, and the DC–DC converter takes care of the current control. A high efficiency can be reached for the latter solution if the nominal power is low. It is advisable to operate the grid-connected inverter in PWM mode if the nominal power is high. The only task for each DC–DC converter is MPPT and perhaps voltage amplification. The DC–DC converters are connected to the DC link of a common DC–AC inverter, which takes care of the grid current control. This is beneficial since better control of each PV module/string is achieved and that common DC–AC inverter may be based on standard VSD technology. (See Figures 12.21 and 12.22 for depictions of a dual-stage inverter and a multistring inverter, respectively.)

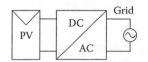

FIGURE 12.20
Single-stage inverter.

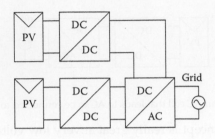

FIGURE 12.21
Dual-stage inverter.

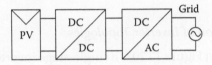

FIGURE 12.22
Multistring inverter.

$$C = \frac{P_{PV}}{2 \cdot \omega_{grid} \cdot U_C \cdot \hat{u}_C} \quad (12.9)$$

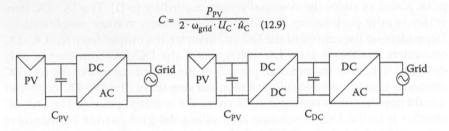

FIGURE 12.23
Filtering for inverter configurations.

12.13.7.4 Power Decoupling

Power decoupling is normally achieved by means of an electrolytic capacitor. As stated earlier, this component is the main limiting factor of the lifetime. Thus, it should be kept as small as possible and preferably substituted with film capacitors. The capacitor is either placed in parallel with the PV modules or in the DC link between the inverter stages; this is illustrated in Figure 12.23.

12.13.7.5 Capacitors

The size of the decoupling capacitor can be expressed as (12.9) where P_{pv} is the nominal power of the PV modules, U_c is the mean voltage across the capacitor, and u_c is the amplitude of the ripple. Equation (12.9) is based on the fact that the current from the PV modules is a pure DC, and that the current drawn from the grid-connected inverter follows a waveform sinus squared, assuming that is constant.

$$C_{PV} = 2,4 \text{ miliF}$$

$$C_{DC} = 33 \text{ microF}$$

Bibliography

1. 2009 Annual World Solar PV Industry Report from Market Buzz.
2. Go Solar California, The California Solar Initiative (http://www.gosolarcalifornia.ca.gov/csi/index.html).
3. Soeren Baekhoej Kjaer, John K. Pederson, and Freda Blaaberg, A review of single-phase grid-connected inverters for photovoltaic modules, *IEEE Transactions on Industrial Applications* 41(5), 2005, 1292–1306.
4. Khomfoi, Surin, and Tolbert, Leon M., Chapter 31, Multilevel power converters, 2008, The University of Tennessee.
5. Wilk, Heinrich, et al., Innovative electrical concepts, Report, International Energy Agency Photovoltaic Power Systems Program, IEA-PVPS 7-07, 2002.
6. Carrasco, Juan Manuel, et al., Power-electronic systems for the grid integration of renewable energy sources: A survey, *IEEE Transactions on Industrial Electronics* 53(4), 2006, 1002–1016.
7. Lai and Peng, TPEL 32(3), 1996.
8. IGCT Devices—Applications and future opportunities, Peter Steimer, Oscar Apeldoorn, Eric Carroll, IEEE PES, Seattle, WA, July 2000.
9. Based on J. M. Carrasco, J. T. Bialasiewicz, et al., Power-electronic systems for the grid integration of renewable energy sources: A survey, *IEEE Transactions on Industrial Electronics* 53(4), August 2006.
10. EN61000-3-2, IEEE1547.
11. U.S. National Electrical Code (NEC) 690.
12. IEC61727.
13. Zipp, Kathleen, Where microinverter and panel manufacturer meet up, Solar Power World, October 24, 2011.
14. Market and technology competition increases as solar inverter demand peaks, Greentech Media Staff from GTM Research. GreentechMedia, May 26, 2009. Retrieved on April 4, 2012.
15. Wesoff, Eric, Solar bridge and PV microinverter reliability, GreentechMedia, June 2, 2011. Retrieved on April 4, 2012.
16. Emerging Renewables Program (ERP) Final Guidebook, February 2009, CEC-300-2009-002-F.
17. California Energy Commission (CEC), Standard for safety inverters, converters, controllers and interconnection system equipment for use with distributed energy resources.
18. Underwriters Laboratories (UL) UL 1741. Note: CEC ERP includes UL 1741.

19. The Encyclopedia of Alternative Energy and Sustainable Living (http://www. daviddarling.info/encyclopedia/S/AE_synchronous_inverter.html Synchronous Inverters).

20. US4362950-1 Synchronous Inverter Compatible with Commerical Power, Dec. 7, 1982.

21. Solar Energy International, *Photovoltaics: Design and installation manual*, New Society, Gabriola Island, BC, Canada, 2006, p. 80.

22. Summary report on the DOE high-tech inverter workshop, Sponsored by the US Department of Energy, prepared by McNeil Technologies (http://eere.energy. gov), accessed June 10, 2011.

23. Go Solar California, List of eligible inverters (gosolarcalifornia.org), accessed July 30, 2009.

Chapter 12 Problems

1. Meteorological data

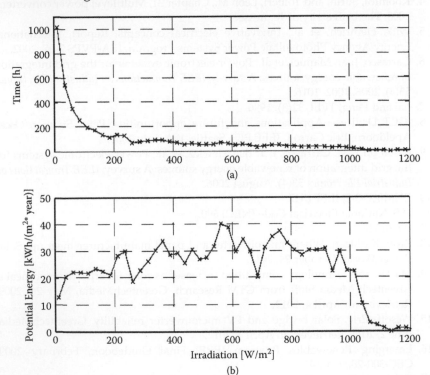

(a)

(b)

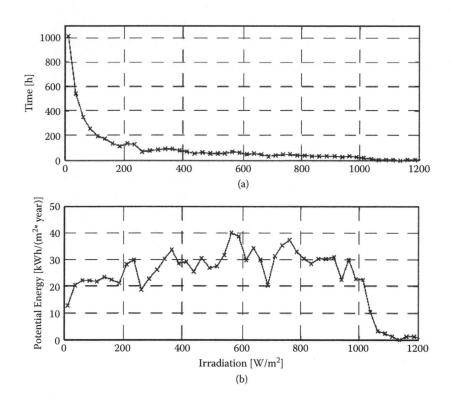

(a)

(b)

2. Find

   a.   Irradiation distribution for a reference year

   b.   Solar energy distribution for a reference year

        Total time of irradiation equals 4686 h per year.

        Total potential energy is equal to 1150 kWh = (m^2 year) 130 W/m^2.

13

Magnetic Inrush Current in Distributed Photovoltaic Grid Power Transformers

Magnetic inrush currents in distributed photovoltaic grid transformers (DPV-GTs) can be as high as 10 to 20 times the nominal current. This is primarily driven by the inductive nature of the transformer coils on the HV and LV sides. The approximate formula for peak inrush current is given by

$$I_{\text{max peak}} = (1000 \times L \times B_s)/(5.2 \times N) \text{ in Amps} \tag{13.1}$$

where
B_s = flux density in air space = $(A_c/A_s)(B \text{ residual} + 2 \times B_{\text{max}} = 130 \text{ kL/in}^2)$ (13.2)
A_c = Area of core in square inches
A_s = Area of air space (core to winding) in square inches + core area in square inches
N = Number of turns in winding (solenoid length) inches
B residual in variable = assume 30 kL/in^2 (i.e., 5 kL/cm^2)
Saturation of iron core assumed at 130 kL/in^2 (i.e., 20.2 kL/cm^2)

Some peak magnetic inrush current curves are shown in Figure 13.1.

13.1 Transformer Inrush Current Protection

- A transformer draws inrush current that can exceed saturation current at power-up (Figure 13.2).
- The inrush current affects the magnetic property of the core.
- This happens even if the transformer has no load with its secondary open.
- The magnitude of the inrush current depends on the point on the AC wave at which the transformer is switched on.
- If turn-on occurs when the AC voltage wave is at its peak value, there will be no inrush current drawn by the transformer. The magnitude of the current in this case will be at normal, no load value.

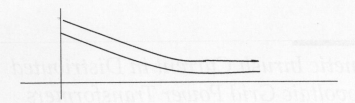

FIGURE 13.1
Peak inrush current curves.

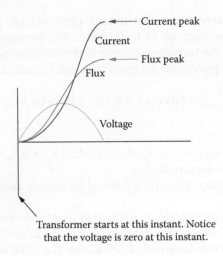

Current peak

Current

Flux peak

Flux

Voltage

Transformer starts at this instant. Notice
that the voltage is zero at this instant.

FIGURE 13.2
(See color insert) Peak current curves for transformer protection.

- If at turn-on the AC wave is going through its zero value, then the current drawn will be very high and exceed the saturation current (see Figure 13.1). In this scenario, the transformer has to be protected from inrush current (see Table 13.1).

13.2 Protection of the Transformer

This application note provides a convenient solution (Figure 13.3) to deal with the problem of inrush current exceeding saturation current in transformers. The solution uses a typical thermistor in series with the primary. This thermistor offers high resistance at the beginning of switching and limits the inrush current. After a short time, the thermistor resistance decreases to a low value due to self-heating and does not affect normal operation.

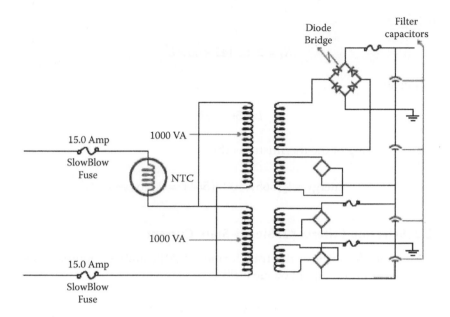

FIGURE 13.3
(See color insert) Typical circuit using a thermistor for protection.

Consider the following DPV-GT:

Each transformer rating: 1000 VA, transformer step-down: 30 V
Total transformer rating: 2000 VA
Filter capacitors used: 30 V, 2300 µF

13.2.1 Selection Criteria #1: Energy

Energy required for the thermistor is computed as follows. We start by calculating the inductive reactance of the transformer:

$$X_L = \text{voltage/Peak Current} = 120/564 = 0.213 \ \Omega$$

Note the following:

- Peak inrush current occurs in one cycle = 564 A, as measured on the oscilloscope
- Input voltage = 120 VAC
- Frequency = 60 Hz

$$X_L = 2\Pi f L$$

Thus,

$$X_L = 2 \times 3.142 \times 60 \times L$$

which yields,

$$L = 565 \ \mu H$$

The energy rating of the thermistor is thus

$$\text{Energy} = \frac{1}{2}\left(565 \times 10^{-6}\right)(564)^2 = 90 \ \text{Joules}$$

13.2.2 Selection Criteria #2: Steady-State Current

Assume that efficiency of the transformer is 70%, ambient temperature is 75°C, and minimum input voltage is 90 V:

$$I_{\text{steady}} = \frac{(\text{KVA of a transformer})}{\left[(\text{Efficiency of Transformer}) \times (\text{Minimum Input Voltage})\right]}$$

For the transformer under consideration,

$$I_{\text{steady}} = (2.0 \ \text{kVA}) / \left[(0.7) \times (90 \ \text{V})\right] = 31.75 \ \text{A}$$

Normally thermistors are rated up to 65°C for their operating current, and then a de-rating factor must be taken into account. Choose an appropriate thermistor that can provide at least the steady-state current as calculated above.

Using the de-rating curve at 75°C and using a corresponding 90% of max rated steady-state current,

$$= 0.90 \times 36 \ \text{A} = 32.40 \ \text{A}$$

One can use any of the thermistors from Tables 13.1 and 13.2 which are rated up to 36.0 A to meet the steady-state current and energy requirements.

13.3 Magnetic Inrush Currents due to Geomagnetic-Induced Currents (GICs)

During a solar storm, as the coronal mass ejection (CME) plasma cloud collides with the planet, large, transient, magnetic perturbations overlay and alter the

TABLE 13.1

Transformer Protection Guide—Typical Inrush Current Limiters for Select Transformers

Transformer KVA	Single-Phase Input Voltage Vac	Continuous Current A	Inrush Current A	Impedance X (Ω)	Inductance Xl (μH)	F (HZ)	Energy (J)	Min R (Ω)
0.50	120	4.16	104	1.63	4328	60	23.4	4.9
1.0	240	4.16	104	3.26	8642	60	46.7	9.78
2.0	240	8.33	208	1.63	4328	60	93.62	4.89
3.0	240	12.5	312	1.09	2881	60	140.6	3.26
5.0	480	10.42	260	2.6	6913	60	234	7.83
10.0	480	20.83	521	1.3	3457	60	469	3.92

TABLE 13.2

Some Inrush Current Limiters for Transformer Applications from Some Manufacturers

Part	UL	R	SSI Max	Joules Max	Voltage Max	Digikey	Mouser	Farnell
SL12 10006	Y	10.0	6	40	240	570-1078-ND	995-SL12-10006	72J6734
SL22 10008	Y	10.0	8	90	240	570-1034-ND	995-SL22-10008	72J6819
SL32 10015	Y	10.0	15	150	240	570-1058-ND	995-SL32-10015	72J6844
AS32 5R020	Y	5.0	20	300	240	570-1106-ND	995-AS32-5R020	
MS32 10015	Y	10.0	15	250	480	570-1014-ND	995-MS32-10015	9006052
MS32 2R025	Y	2.0	25	300	480	570-1019-ND	995-MS32-2R025	72J6622
MS35 5R025	N	5.0	25	600	680	570-1029-ND		72J6634

normally stable magnetic field of Earth. These magnetic perturbations are referred to as a *geomagnetic storm* and can affect the planet for a period of a day or two. These perturbations can induce voltage variations along the surface of the planet and induce electric fields in the Earth to create potential differences in voltage between grounding points, which causes geomagnetic-induced currents (GICs) to flow through transformers, power system lines, and grounding points. GICs can severely affect grounded wye-connected transformers and autotransformers through its Earth-neutral connection.

GICs can cause transformers to be driven into half-cycle saturation where the core of the transformer is magnetically saturated on alternate half-cycles. Only a few amperes are needed to disrupt transformer operations. A GIC level induced voltage of 1 to 2 V/km and 5 A in the neutral of the high-voltage windings is sufficient to drive grounded wye-connected distribution transformers into saturation in a second or less.

During geomagnetic storms, GIC currents as high as 184 A have been measured in the United States in the neutral leg of transformers.

The largest GIC measured thus far was 270 A during a geomagnetic storm in Southern Sweden on April 6, 2000. If transformer half-cycle saturation is allowed to continue, stray flux can enter the transformer structural tank members and current windings. Localized hot spots can develop quickly inside the transformer tank as temperatures rise hundreds of degrees within a few minutes.

Temperature spikes as high as 750°F have been measured. As transformers switch 60 times per second between being saturated and unsaturated, the normal hum of a transformer becomes a raucous, cracking whine. Regions of opposed magnetism as big as a fist in the core steel plates crash about and vibrate the 100-ton transformers that are nearly the size of a small house. This punishment can go on for hours for the duration of the geomagnetic storm. GIC-induced saturation can also cause excessive gas evolution within transformers. Besides outright failure, the evidence of distress is increased gas content in transformer oil, especially those gases generated by decomposition of cellulose, vibration of the transformer tank and core, and increased noise levels of the transformers (noise level increases of 80 dB have been observed).

GIC transformer damage is progressive in nature. Accumulated overheating damage results in a shortening of the transformer winding insulation life span, eventually leading to premature failure.

In addition to problems in the transformer, half-cycle saturation causes the transformer to draw a large exciting current that has a fundamental frequency component that lags the supply voltage by 90° and leads to the transformer becoming an unexpected inductive load on the system. This results in harmonic distortions and added loads due to reactive power or volt-ampere reactive (VAR) demands. This results in both a reduction in the electrical system voltage and the overloading of long transmission tie-lines. In addition, harmonics can cause protective relays to operate improperly and shunt capacitor banks to overload. The conditions can lead to major power failures. When induced current flows into the electrical power grid, it has the potential of overloading the grid and causing significant damage to critical components at power plants. Solar storms can cause major electrical blackouts that can affect millions of people over a large geographic area. The electrical power grid in the United States is composed of several elements. The electricity is generated in hydroelectric dams, coal-/gas-/oil-fired power plants, and nuclear power plants. The backbone of the electrical power grid is formed by the high-voltage transmission lines operating at 230, 345, 500, and 765 kV. These transmission lines and their associated transformers serve as the long-distance heavy-hauling arteries of electricity in the United States. The electricity is transferred between power generators and regional substations on very heavy supply lines suspended on 100-foot-tall towers. These cables generally are three-phased systems with two hot lines and one ground. At the regional substations the voltage is converted into lower voltages from 69,000 to 13,800 V. The substations feed local communities generally using telephone poles. Individual or neighborhood transformers step the voltage down to 220 V which supplies homes and businesses with electrical power. The U.S. electrical system includes over 6,000 generating units, more than 500,000 miles of bulk transmission lines, approximately 12,000 major substations, and innumerable lower-voltage distribution transformers. All can serve as potential GIC entry points from their respective ground connections. This enormous network is controlled regionally by more than 100 separate control centers that coordinate responsibilities jointly for the impacts upon real-time network operations.

14

Eddy Current and Stray Loss Calculations of Distributed Photovoltaic Grid Power Transformer

Eddy current losses (ECLs) and stray losses are present in every transformer. The primary stray and eddy losses are due to the 60 Hz frequency currents and related thickness of the conductor thickness of the windings. These loss components increase with the square of the frequency and square of the magnitude of the eddy currents. If the inverter feeding the power into the step-up transformer is producing more than the standard level of harmonics, then the stray and eddy losses will increase. The increase in load loss effect on efficiency is not typically a concern. Of much greater concern is the increased hot-spot temperature in the windings that can reduce transformer life. A specially designed transformer can compensate for the higher stray and eddy losses. Also, a larger than necessary kVA transformer can be selected to compensate for the higher operating temperatures. However, these concerns on ECL increasing are generally mitigated because harmonics are less than 1%. However, because DPV-GTs have high-current windings on the low-voltage side, the conductor thickness becomes a focal point in accentuating the effects of additional ECL as described below.

14.1 Eddy Current Loss (ECL) in DPV-GT

The eddy current loss (ECL) [1,2,4] in Figure 14.1 is given by

$$W = k_2 f^2 \left(B_{\text{eff}}\right)^2 t^2 \text{ w/lb}$$

$$= k_4 \left(B_{\text{eff}}\right)^2 t^2 \text{ w/lb} \qquad (14.1)$$

where B_{eff} is the nominal flux density, E_{eff} is the root mean square (rms) voltage, t is the thickness of the lamination, k_4 is a constant depending on the material, and f is the frequency of the power signal.

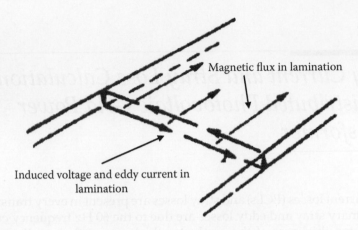

Magnetic flux in lamination

Induced voltage and eddy current in
lamination

FIGURE 14.1
Sinusoidally varying magnetic flux induces a current in the surface of the lamination called
the eddy current.

ECL in the core has a fairly high power factor resistance loss proportional
to the square of the rms value of the excitation voltage, independent of the
shape of the wave. If this voltage is held in accordance with the average voltage measured by a voltmeter (i.e., actual rms value not to be read value), the
observed ECL will be a multiple or a fraction of the true ECL by a factor K.

The correction in the ECL factor is as follows:

$$K = (\text{Actual Eddy Loss})/(\text{Eddy Loss for sine wave voltage})$$

$$= \{(\text{actual rms voltage})/(\text{average voltage} \times 1.1)\}^2$$

$$= \{(\text{actual rms voltage})/(\text{Rated rms voltage})\}^2 \qquad (14.2)$$

$$\text{Total sine wave loss} = \text{Observed loss} \times \{(100)/(\%\text{Hys} + \%\text{ECL})\} \quad (14.3)$$

14.2 Alternate Method for ECL in Windings

Let A be the bare width of the conductor, B be the covered width of the
conductor plus block thickness, H be the bare thickness of the conductor in
centimeters, and F be the frequency of the power supply. Then

$$\alpha = \text{Sqrt}\left(\frac{\dfrac{F}{50} \times \dfrac{2}{B} \times \dfrac{47.8}{50}}{1}\right) \qquad (14.4)$$

TABLE 14.1

Values of ε

ε	$Q(\varepsilon)$	$\Psi(\varepsilon)$
0.0	1.00	0
0.1	1.0	0.0000332
0.2	1.0	0.00053
0.3	1.001	0.0027
0.4	1.002	0.0085
0.5	1.005	0.0207

Note: $\Psi(\varepsilon)$ varies as the fourth power of ε. Hence, intermediate values can be found using Equation (14.5).

and we evaluate

$$\varepsilon = \alpha \times H \qquad (14.5)$$

$$Average\left(Rac @ 75°C / Rdc @ 75°C\right) = Q(\varepsilon) + \left\{ \left(m^2 - 1\right) / (3) \right\} \cdot \psi(\varepsilon) \quad (14.6)$$

(See Table 14.1.) This yields ECL as a % of

$$I^2 Rdc = \left\{ Q(\varepsilon) + \left[\left(m^2 - 1\right) / (3) \right] \cdot \psi(\varepsilon) \right\} \times 100 \qquad (14.7)$$

where m is the number of conductors in the radial direction.

All of the dimensions used in Equation (14.7) are either in inches or millimeters unless otherwise specified.

14.3 Calculation of Stray Losses

Consider the following values in the calculation of stray losses: d_{lv} = radial depth of LV winding; d_{hv} = radial depth of HV winding; d_{hl} = radial duct between HV and LV winding; MDh_l = mean diameter of HV-LV gap; and K = constant value to be taken from stray-loss curve, generally 35. (See Figure 14.2.) Thus,

$$Stray\ Loss = K \left\{ \frac{\left(\text{Ampere turns at normal tap}\right)}{\left(\text{Leg length of Core} \times 1000\right)} \right\}^2$$

$$\times \left\{ \left(d_{lv} + d_{hv}\right) / 3 + d_{hl} \right\} \times MDh_l \qquad (14.8)$$

For all of the above calculations, the dimensions have to be in inches or centimeters unless otherwise specified.

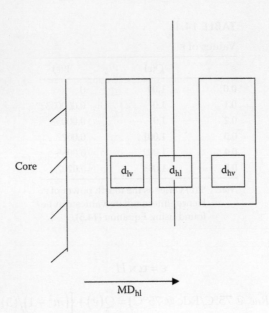

FIGURE 14.2
Core-coil disposition of a two-winding DPV-GT.

15

Design Considerations—Inside/Outside Windings for a Distributed Photovoltaic Grid Power Transformer

15.1 Design of Windings

Windings form the most important part of a transformer, which is necessary to affect the required transfer of electric power from the primary or source side to the secondary or load side at maximum efficiency. These issues become more critical while designing transformers suitable for solar energy applications because of the sudden changes in various factors like voltage variations, harmonics, and frequency variation in general.

The design of the windings of DPV-GTs depends on the intention of the designer to satisfy the following salient features:

- Ensure adequate effective dielectric strength suitable to withstand
  - Operating voltages and their changes for solar transformers
  - System fault voltages during single line to ground (SLG) fault; double line to ground (DLG) fault, and all lines to ground (ALG) faults
  - Switching surges caused by separated microchanges in the requirements from *prosumers*. This is a new term coined recently after the advent of smart grids and the integration of microgrids formed using alternate energy sources. The solar energy penetration in this area is constantly increasing.
  - Lightning surges
  - Test voltages; IEEE and IEC test considerations have become increasingly stringent. In addition, since private agencies are now involved in the creation of smart grids, much more stringent test conditions appear on the horizon. For example, the partial discharge test considerations for Class I transformers call for a very relaxed pico Coulomb level, while end users who dictate the terms for their network transformers may have lower values of the partial discharge than those prescribed by IEEE.

- Maintain rated temperature rises within the winding system by providing adequate coil ventilation. With the reduction in overall physical sizes of transformers and the revolution in the design process,
  the physical clearances in all core-coil assembly dispositions have
  reduced many fold. These demand much better and efficient cooling
  systems for all insulation components in such transformers for solar
  energy application.

- Possess sufficient mechanical strength to withstand forces created
  by opposing currents in the HV and LV windings, especially in the
  multiwinding designs.

- Use a minimum amount of winding material to optimize the cost
  of the transformer as a system device with a current density not to
  exceed 2.2 A/mm^2.

- Consider that a system can have multiwinding or multicircuit
  arrangements to achieve impedance requirements specified by the
  user for up to 18–20%.

- Minimize the eddy current loss (ECL).

- Take into account partial discharge considerations per IEEE C57.127.

15.2 Kinds of Windings

There are two main kinds of windings: concentric and pancake (or round).
Properties associated with them are as follows:

- High-voltage coils: High-voltage coils are the components of finished transformers. They are made on automatic layer setting winding machines.

- A solid cylindrical form of predetermined diameter and length is being
  used as a base over which a winding is wound. If multiple windings
  exist, they are concentric and wound on top of the inner windings
  with sufficient cooling and voltage-withstanding insulation.

- Generally round insulated wire of either copper (Cu) or aluminum
  (Al) is used as the basic raw material.

- The coils are made in a number of layers for layer-type winding or
  helical winding.

- The starting and finishing leads of each coil are terminated on either
  side of the coil.

- These leads are properly sleeved and locked at a number of points.

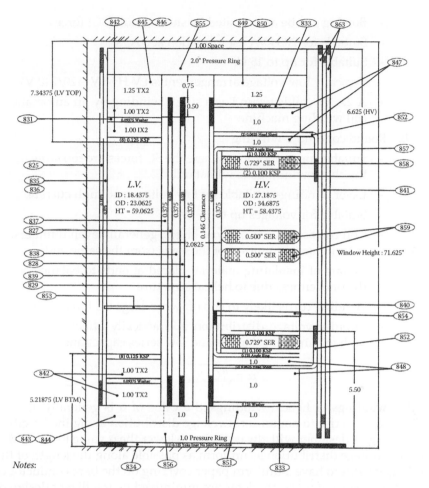

FIGURE 15.1
Typical core-coil assembly for DPV-GT.

- The shape of the basic raw material (Al or Cu) is rectangular. For small transformers, these shapes could be round as well. The former shape will increase the manufacturing of rectangular shape windings with automated machines as shown in Figure 15.1.

Concentric windings

- HV wound over LV
  a. Rectangular concentric winding
     – Very little space is wasted
     – Can cause excessive bowing of the conductors away from flat surfaces

- Bowing can be accentuated by short circuit (SC) forces
- Suitable for 200 kVA leg formation at 60 c/s
- Suitable for up to 15 kV
- Typically low voltage arrangement is LV-Hv-LV (120/240 V)
- A 2.4 kV coil is wound as a single winding by an automatic coil winding machine

b. Round concentric winding

- Suitable for applications where the SC forces are too large for the above windings to withstand
- Layer winding—suitable for low voltage and high current
- Suitable for voltages up to 15 kV
- Coil consists of one or more layers, generally even numbers, to facilitate the removal of the leads from the same end
- Collars of insulating material placed at ends to balance out the unevenness due to helical structure
- Winding turns likely to slip off when large SC forces exist
- Impulse voltage distribution is drastically affected due to high ground capacitances and low series capacitance

$$\alpha = \sqrt{\left(C_g / C_s\right)} \tag{15.1}$$

For disc windings or helical windings, the value of α is generally about 2 to 2.5. This gives rise to nonuniform voltage distribution along the length of the winding. A designer will always strive to lower the α value to near 1.0 to make the voltage distribution as uniform as possible along the length of the winding. It helps to have a uniform paper covering on the bare conductor of the winding. Most of these problems are minimized by intelligent design of windings. One such way to make the voltage distribution along the length of the winding uniform is by using low series capacitor windings. In some voltage ratings as high as 66 kV, the interleaved winding structure is justifiable. A typical core-coil assembly of a DPV-GT is shown in Figure 15.1. Figure 15.2 shows the winding of a rectangular form of winding using an automatic foil winding machine. Figure 15.3 shows the spools that carry the raw material for the windings to be automatically wound by the machine.

Table 15.1 shows the typical sizes of aluminum conductors with the paper covering (Pc) and the weights for DPV-GTs up to 1000 kVA.

Tables 15.1 and 15.2 are used to choose the proper conductor for different rating transformers. The current density in the conductors is generally kept at 2.2 A/mm². Copper or aluminum conductors are generally obtained from manufacturers of conductors in spools called *drums*. Sometimes because of the value of the current, multiple conductors are placed in a bunch. To minimize the difference in the lengths of each of the conductors in the bundle, the

FIGURE 15.2
(See color insert) Foil winding machine with foil being wound.

FIGURE 15.3
(See color insert) Spools, which carry the aluminum foil, assembled at the back of the foil winding machine.

conductors are periodically transposed, thus also called *continually transposed conductors* (CTCs).

15.2.1 Pretransposed Strip Conductor

Enameled conductor proportions: minimum size = 3.0 × 1.0 mm

Maximum size = 11.0 × 3.0 mm

Width/thickness: ratio between limits (5:1) & (2:1)

TABLE 15.1

Conductor Sizes and Quantities for Standard Aluminum Wound Transformers

KVA	Losses (W) and % Z No load/Load/Z	HV Conductor 11 kV Size	Wt kg	6.6 kV Size	Wt kg	LV Conductor 400 V Size mm	Wt kg	416–420 V Size mm	Wt kg	433–440 V Size mm	Wt kg
160	550/3050/4.75	0.068″ Pc 12	42	0.086″ Pc 12	39	14 × 4 Pc 0.4	25	14 × 4 Pc 0.4	27	14 × 4 Pc 0.4	27
250	700/4350/4.75	0.086″ Pc 12	56	0.110″ Pc 12	56	10 × 4 Pc 0.4	32	10 × 4 Pc 0.4	34	10 × 4 Pc 0.4	35
315	780/5700/4.75	0.092″ Pc 12	62	0.116″ Pc 12	59	14 × 4 Pc 0.4	20	14 × 4 Pc 0.4	21	14 × 4 Pc 0.4	21
						10 × 4 Pc 0.4	14	10 × 4 Pc 0.4	15	10 × 4 Pc 0.4	15
400	890/6600/4.75	0.110″ Pc 12	80	0.136″ Pc 14	73	10 × 4 Pc 0.4	41	10 × 4 Pc 0.4	43	10 × 4 Pc 0.4	44
500	1080/7800/4.75	0.116″ Pc 12	81	0.164″ Pc 14	100	11 × 4.5 Pc 0.4	49	11 × 4.5 Pc 0.4	51	11 × 4.5 Pc 0.4	52
630	1500/8900/4.75	0.136″ Pc 14	110	0.164″ Pc 14	98	11 × 4.5 Pc 0.4	63	11 × 4.5 Pc 0.4	66	11 × 4.5 Pc 0.4	68
750	1390/10100/5.0	0.144″ Pc 14	115	5.5 × 3.5 Pc 0.4	131	12 × 5 Pc 0.4	74	12 × 5 Pc 0.4	77	12 × 5 Pc 0.4	80
1000	1600/13300/5.0	0.164″ Pc 14	128	6.5 × 4 Pc 0.4	142	11 × 4.5 Pc 0.4	73	11 × 4.5 Pc 0.4	77	11 × 4.5 Pc 0.4	80
				copper		copper		copper		copper	

TABLE 15.2

Standards of Reference for Enameled Conductor and Paper Insulation

Drum Size	Outside Flange Diameter	Barrel Diameter	Overall Width	Transverse	Hole Center	Driving	Flange Depth[a]	Flange Width
Standard 7-foot drum	2130	1830	1020	810	78	57	140	76
Standard 5-foot (A type)	1520	1220	970	810	78	57	140	76
Standard 5-foot (B type)	1520	1070	970	810	78	57	216	76
Standard 5-foot (narrow)	1520	1140	470	340	78	57	178	64

[a] Flange depth, number of layers of transposed strip that may be accommodated in drum.

15.2.2 Transposed Conductor Proportions

Number of individual conductors = 5 (minimum)

= 31 (maximum)

The conductor stack proportions (under paper) should preferably be kept within the following limits:

Radial height/axial width

1 : 2½   (small number of strip)

3 : 1   (large number of strip)

Dimensions:

Individual conductors are covered with m-grade enamel covering 0.1 mm.

Axial width (mm) = 2 × (bare width of each conr + 0.1)

+ overall paper covering + 0.4

Radial height (mm) = [(n+1)/2] × (bare height of each conductor + 0.1)

+ overall paper covering + 0.8

Length and weight of conductor:

Not to exceed 1500/2000 m and 3000 kg

FIGURE 15.4
(See color insert) Rectangular aluminum foil, layer or helitran winding.

Frequency of transposition:

Pitch of transposition varies between 64 and 203 mm according to size

Bending properties:

Winding diameter should not be less than the value given by *dia* = 0.6 *nw*, where *n* = number of conductors, and *w* = axial width of conductor.

DPV-GTs used as industrial transformers are easy to manufacture with windings in a rectangular form for compactness (Figure 15.4). Automatic foil winding machines are used to wind the low-voltage winding with aluminum. The high-voltage windings are then wound right on top of the low-voltage windings with suitable insulation, generally obtained in sheet form with strips pre-glued on to them. These transformers also have rectangular cross-sectional magnetic cores. Thus it is easy to increase the productivity of a production shop by using these machines for small transformers up to 10 MVA, 33 kV. Recently this rating has been increased to 20 MVA, 69 kV by the use of improved insulation material and compact designs (Tables 15.3 through 15.7). The most intriguing use of this method is in the manufacture of windings for transformers using superconductors.

The conductors are used in a stack typically for LV winding to share the high current, and a mechanism called *transposition* is used as described below. This ensures that the total length of all the conductors in the stack

TABLE 15.3

Conductor Sizes and Quantities for Standard Copper Wound Transformers

KVA	Losses (W) and % Z No load/Load/Z	HV Conductor 11 kV Size	Wt kg	6.6 kV Size	Wt kg	LV Conductor 400 V Size mm	Wt kg	416–420 V Size mm	Wt kg	433–440 V Size mm	Wt kg
315	810/5000/4.75	0.072" Pc 12	98	0.092" Pc 12	98	10 × 3.5 Pc 0.4	76	10 × 3.5 / 9 × 3.5 Pc 0.4	41 / 37	9 × 3.5 Pc 0.4	75
400	970/5900/4.75	0.080" Pc 12	113	0.104" Pc 12	116	10 × 3.5 / 12 × 3.5 Pc 0.4	36 / 44	10 × 3.5 / 12 × 3.5 Pc 0.4	38 / 45	9 × 3.5 / 12 × 3.5 Pc 0.4	35 / 47
500	1220/6900/4.75	0.092" Pc 12	139	0.116" Pc 12	132	9 × 3.5 Pc 0.4	90	9 × 3.5 Pc 0.4	94	9 × 3.5 Pc 0.4	98
630	1280/8600/4.75	0.104" Pc 12	172	0.128" Pc 12	152	12 × 3.5 Pc 0.4	119	12 × 3.5 / 10 × 3.5 Pc 0.4	83 / 35	10 × 4 / 9 × 3.5 Pc 0.4	86 / 32
750	1550/9450/5.0	0.110" Pc 12	179	0.144" Pc 14	183	11 × 4.5 Pc 0.4	131	11 × 4.5 Pc 0.4	135	11 × 4.5 Pc 0.4	142
1000	1780/120000/5.0	0.128" Pc 12	227	0.164" Pc 14	231	11 × 3 Pc 0.4	167	11 × 3 Pc 0.4	172	11 × 3 Pc 0.4	180
1250	2060/14100/5.0	0.144" Pc 14	284	7.5 × 2.4 Pc 0.4	265	11 × 3 Pc 0.4	201	11 × 3 Pc 0.4	201	11 × 3 / 10 × 3 Pc 0.4	128 / 78
1500	2450/17000/5.0	0.164" Pc 14	334	6.6 × 3.6 Pc 0.4	332	10 × 3.5 / 10 × 3 Pc 0.4	160 / 69	10 × 3.5 / 10 × 3 Pc 0.4	82 / 141	10 × 3 Pc 0.4	220
1600	2450/17800/5.0	0.164" Pc 14	334	6.6 × 3.6 Pc 0.4	332	10 × 3.5 Pc 0.4	237	10 × 3.5 Pc 0.4	254	10 × 3.5 Pc 0.4	254

TABLE 15.4

Standard Conductor Sizes

Copper		Aluminum	
Round (inches)	Strip (mm)	Round (inches)	Strip (mm)
0.072" pc 12	9 × 3.5 pc 0.4	0.068" pc 12	10 × 4.0 pc 0.4
0.080" pc 12	10 × 3.0 pc 0.4	0.086" pc 12	11 × 4.5 pc 0.4
0.092" pc 12	10 × 3.5 pc 0.4	0.092" pc 12	12 × 5 pc 0.4
0.104" pc 12	11 × 3.0 pc 0.4	0.110" pc 12	14 × 4 pc 0.4
0.110" pc 12	11 × 4.5 pc 0.4	0.116" pc 12	5.5 × 3.5 pc 0.4
0.116" pc 12	12 × 3.5 pc 0.4	0.136" pc 14	6.5 × 4 pc 0.4
0.128" pc 12	6.6 × 3.6 pc 0.4	0.144" pc 14	
0.144" pc 14	7.5 × 2.4 pc 0.4	0.164" pc 14	
0.164" pc 14			

Note: Ratings covered: (a) Copper: 315, 400, 500, 630, 750, 1000, 1250, 1500, and 1600 kVA; (b) Aluminum: 160, 250, 315, 500, 630, 750 both HV and LV; 1000 KVA, HV aluminum, LV copper.

TABLE 15.5

Copper Conductors for DPV-GTs

KVA	Winding	Conductor Size (round in inches, strip in mm)
500	HV	0.092" pc 12
500	LV	9 × 3.5 pc 0.4
1000	HV	0.128" pc 12
1000	LV	11 × 3 pc 0.4
1000	LV	11 × 4.5 pc 0.4
1500/1600	HV	0.164" pc 14
1500/1600	LV	10 × 3 pc 0.4
1500/1600	LV	10 × 3.5 pc 0.4

in parallel remains the same to yield uniform resistances in parallel. Some transposition schemes for odd and even numbers of conductors in a stack are illustrated below.

15.2.3 Rotary Transposition for Helical or Spiral Windings for Core-Type DPV-GTs

Note the positions of the conductors, as these are wound on the former and the subsequent transpositions, thus ensuring equal length of the conductors for each layer of the helical or spiral winding, resulting in uniform resistance of the overall winding conductors. Note the direction of the arrows for proper transposition to be effected (Figures 15.5 and 15.6).

TABLE 15.6

Turn Insulation

Thickness of Turn Insulation (mils)	Turn Insulation Only	
	Maximum rms Power Frequency 1 minute withstand kV	Maximum Momentary Impulse withstand kV
10	4	16
20	7.5	30
30	10.75	43
40	14	56
50	17	68
60	20	80
70	23	92
80	26	104
90	29	116
100	32	128
120	37.5	150
140	43	172
160	48	192
180	53	212
200	58	232
250	70	280
300	81	324
350	91	364
400	100	400
450	108	432
500	115	460

Turn Insulation plus Spacing between Coils	
Ratio: (spacing between coils)/ (thickness of turn insulation)	% of momentary impulse withstand (kV)
1	100
2	120
3	135
4	150
5	160
6	169
7	178
8	186
9	123
10	200
15	225
20	245
30	285
40	315
50	340
75	400
100	450

TABLE 15.7

Common Copper Conductors Available in Market

SWG	Bare Conductor Diameter		Max Overall Covered Diameter (cm)		Conductor Area (cm²)	Copper		Aluminum		
	Inches	Centimeters	Med SECM	Thick SECT		Bare wt. (kg/km)	Resistance (Ohm/km)	Bare wt. (kg/km)	Resistance (Ohm/km)	SECM wt. (kg/km)
8	0.160	4.064	4.221	4.255	12.972	115.32	1.625	35.02	2.6981	0.84
9	0.144	3.658	3.807	3.840	10.507	93.41	2.0081	28.37	3.3311	0.73
10	0.128	3.251	3.396	3.429	8.302	73.80	2.5415	22.42	4.2158	0.65
11	0.116	2.946	3.086	3.119	6.818	60.61	3.0947	18.41	5.1334	0.58
12	0.104	2.642	2.774	2.807	5.481	48.72	3.8496	14.80	6.3856	0.50
13	0.092	2.336	2.464	2.494	4.289	38.13	4.915	11.58	8.1604	0.43
14	0.080	2.032	2.151	2.182	3.243	28.83	6.5063	8.756	10.792	0.36
	0.076	1.930	2.047	2.075	2.927	26.02	7.2080	7.902	11.957	0.34
15	0.072	1.892	1.943	1.971	2.627	23.35	8.0319	7.092	13.323	0.32
	0.068	1.727	1.839	1.867	2.343	20.83	8.9293	6.326	14.938	0.29
16	0.064	1.626	1.725	1.763	2.076	18.45	10.163	5.604	16.859	0.27
	0.060	1.524	1.631	1.659	1.824	16.22	11.657	4.925	19.188	0.24
17	0.056	1.422	1.529	1.557	1.589	14.13	13.278	4.290	22.026	0.22
	0.052	1.321	1.420	1.448	1.370	12.18	15.401	3.699	25.547	0.21

18	0.048	1.219	1.316	1.346	1.168	10.38	18.065	3.152	29.965	0.19
	0.044	1.117	1.212	1.237	0.9810	8.72	21.508	2.649	35.677	0.17
19	0.040	1.016	1.110	1.135	0.8107	7.207	26.026	2.189	43.172	0.154
	0.0380	0.965	1.057	1.079	0.7317	6.500	28.836	1.976	47.833	0.140
20	0.036	0.914	1.006	1.029	0.6567	5.838	32.130	1.773	53.296	0.134
	0.034	0.864	0.947	0.970	0.5858	5.210	36.018	1.582	59.747	0.120
21	0.032	0.813	0.897	0.919	0.5189	4.613	40.662	1.401	67.450	0.110
	0.030	0.762	0.843	0.866	0.4560	0.50	46.271	1.231	76.754	0.100
22	0.028	0.711	0.792	0.815	0.3973	3.532	53.108	1.073	88.094	0.093
	0.026	0.660	0.739	0.759	0.3425	3.045	61.605	0.9248	102.18	0.085
23	0.024	0.610	0.686	0.706	0.2919	2.595	72.285	0.7881	119.90	0.072
	0.023	0.584	0.660	0.681	0.2680	2.383	78.731	0.7236	130.59	0.073
24	0.022	0.559	0.632	0.653	0.2452	2.180	86.052	0.6620	142.74	0.068
25	0.020	0.508	0.579	0.599	0.2027	1.802	104.094	0.5473	172.66	0.061
26	0.018	0.457	0.526	0.546	0.1642	1.460	128.501	0.4433	213.15	0.052
27	0.0172	0.437	0.505	0.526	0.1499	1.333	140.760	0.04047	233.48	0.049
	0.0164	0.417	0.483	0.500	0.1363	1.212	153.678	0.3680	256.78	0.046
28	0.0148	0.376	0.434	0.452	0.1110	0.987	190.009	0.2997	315.31	0.038
29	0.0136	0.345	0.404	0.422	0.0937	0.833	225.186	0.2530	373.53	0.035
30	0.0214	0.315	0.371	0.389	0.0779	0.693	270.186	0.2103	449.29	0.031

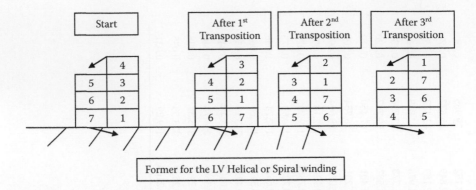

FIGURE 15.5
Transposition for helical or spiral windings of LV with odd conductors in a stack.

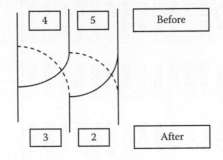

FIGURE 15.6
Plan of strips at first transposition in reference to Figure 15.5.

A similar transposition scheme is illustrated for an even number of conductors in a stack. Note the direction of the arrows that indicates the move for the transposition as the winding is being built on the former for the corresponding LV helical or spiral winding (Figures 15.7 and 15.8).

15.3 Typical Conductor Transposition Examples in Spiral Windings Used in DPV-GTs

Distributed transposition: In this transposition the conductor forming the top layer is taken to the bottom, and all other conductors are moved up one conductor depth as shown in Figure 15.9. For one complete transposition there will be $(n - 1)$ distributed transpositions, where n is the number of conductors in parallel in the radial direction.

Complete transposition: The conductor forming the top layer is taken to the bottom. The first conductor from the top is taken to the first

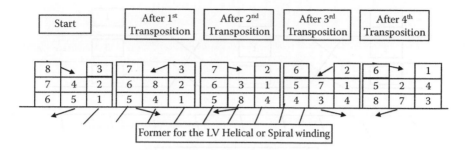

FIGURE 15.7
Transposition for helical or spiral windings of LV with even conductors in a stack.

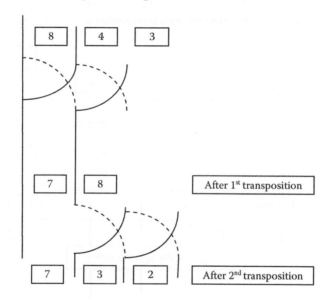

FIGURE 15.8
Plan of strips at first transposition in reference to Figure 15.7.

from the bottom, and so on. If the number of conductors in the radial direction is odd, the center conductor does not change its position. In the length of each layer there will generally be one complete transposition at the center turn (Figure 15.10).

Multiple transpositions:

- At one-fourth coil length, the top half of the conductors are interchanged as a group with the bottom half.
- At half the coil length, a complete transposition (as shown in Figure 15.11) changeover is made.
- At the three-fourth coil length, the top half of the conductors are again interchanged as a group with the bottom half.

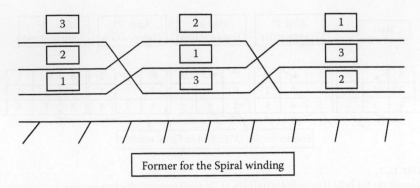

Former for the Spiral winding

FIGURE 15.9
Distributed transposition with $(n - 1)$ distribution transpositions.

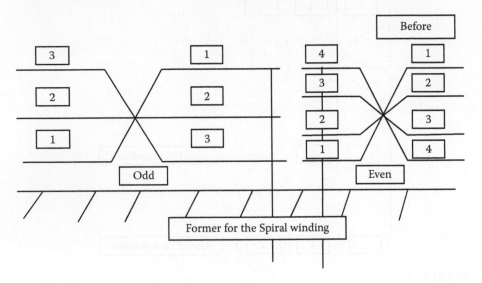

Former for the Spiral winding

FIGURE 15.10
Complete transposition of conductors in a spiral winding stack.

All conductors in the multiple conductor spiral windings are transposed in the radial direction of the windings according to one of the above methods.

15.3.1 Transposition of Bunched Conductors

15.3.1.1 Stage 1

Before transposition, conductors are cut at position *D*. The common insulation covering is removed between points *A* and *B*. The individual conductor covering is not to be removed or disturbed or severe failures internal to the transformer may be caused (Figure 15.12).

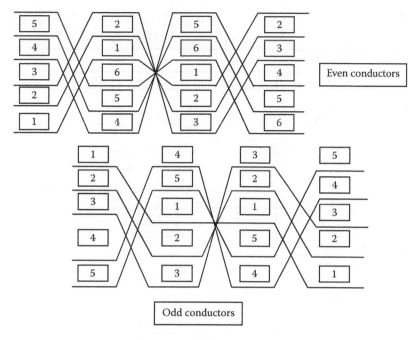

FIGURE 15.11
Multiple transpositions of conductors in a spiral winding stack.

15.3.1.2 Stage 2

Conductors are arranged after transposition as shown in Stage 2 and are butt jointed at D, using a butt welder. After transposition each conductor is taped to the required thickness as required by the voltage rating between points C and D. All conductors are bunched together under common covering behind F and after E. If any conductor is slightly loose, it is taped with a wooden hammer and tied to a disc. In the case of rectangular foil windings, the conductor is a flat sheet and is generally wound on rectangular formers. In such cases, the core is also rectangular in cross section and forms the former for the winding closest to the core. Proper insulation wraps and spacers are used to coordinate the insulation level for each of the windings.

15.3.2 Forces on Windings

According to Ampere's law, by the basic interaction of the currents between the primary and the secondary windings, the equation for force is

$$F = \left[(\mu I_1)/2\pi r \right] [I_2 l] \, ; \text{Newtons} \tag{15.2}$$

$$= I_2 l B \tag{15.3}$$

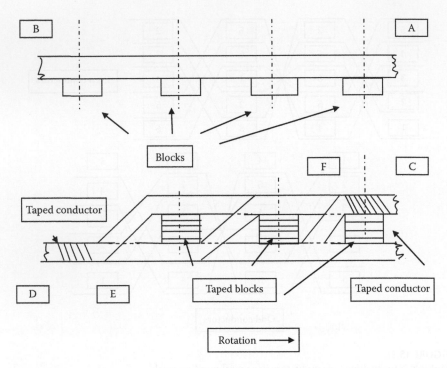

FIGURE 15.12
Transposition of bunched conductors.

where

$$B = (\mu I_1)/(2\pi r); \; \text{Wb}/\text{m}^2 - \text{Magnetic flux density} \tag{15.4}$$

The direction of the magnetic field is decided by the conventional right-hand (three-finger) rule, where the index finger is pointed in the direction of the current, the middle finger in the direction of the force, and the thumb in the direction of the flux density.

Forces between conductors of the primary and secondary windings of a transformer lead to radial and axial forces. The former cause the winding close to the core to compress radially and the winding on the outer diameter to expand radially. The latter forces press the windings upward or downward into the core for a core-type transformer. In the shell-type construction, because the windings are surrounded by the core and the windings are configured in a "pancake" mode, both radial and axial forces will exist.

Thus, mechanical forces on windings (Figure 15.13) in a transformer per unit length of a conductor are

$$F = B \times I \; (\text{current in the conductor}) \tag{15.5}$$

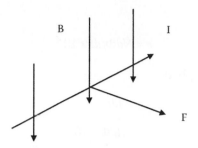

FIGURE 15.13
Mechanical force on windings under the influence of a magnetic field *B*.

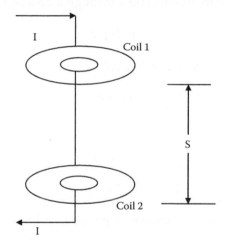

FIGURE 15.14
Forces between two coils.

15.3.3 Forces between Two Coils in Series

Let *I* be the current flowing in coil 1, which is connected in series with coil 2. Let the total inductance change

$$L = L + dL \tag{15.6}$$

when coil 1 moves by a distance dS (Figure 15.14). Energy in the magnetic field changes from

$$1/2 LI^2 \text{ to } 1/2 (L + dL)^2 \tag{15.7}$$

and flux linkage changes from

$$N\Phi/I \text{ to } N(\Phi + d\Phi)/I \tag{15.8}$$

Thus,

$$e = Nd\Phi/dt \times 10^{-8} \tag{15.9}$$

Additional energy supplied

$$= e \; Idt$$

$$= NId\Phi \times 10^{-8}$$

$$= dLI^2 \tag{15.10}$$

Therefore, work done in moving coil 1 through a distance dS is

$$= \left(dLI^2\right)/2 \tag{15.11}$$

Force acting on this dS element of the coil is

$$F = \text{work/distance}$$

$$= \left(dLI^2\right)/2/dS$$

$$= \left(I^2/2\right)\left(dL/dS\right) \text{ dynes} \times 10^7$$

$$= 4.42 \; I^2 \; dL/Ds \tag{15.12}$$

where I is in amps, L is in henry, F is in pounds, and S is in centimeters.

15.3.4 Forces in Concentric Coils

Since the force is perpendicular to the current and flux, principal forces in a conductor of concentric foils are radial. These forces are

$$F\alpha B \text{ and } I \tag{15.13}$$

$$B_{\max} = 0.4\pi \; NI/l \tag{15.14}$$

Thus,

$$F = \left(0.4\pi\sqrt{2I_pN_p} \cdot \sqrt{2I_c}\right)/l \tag{15.15}$$

where F = outward force on one conductor on inside of outer coil, in crest value in dynes per length; I_pN_p = rms ampere turn in primary coil; I_c = rms amperes in conductor; and l = axial length of primary coil in cm:

$$F = \left(5.64 \times I_pN_p \times I_c\right)/\left(10^7 \times L\right) \tag{15.16}$$

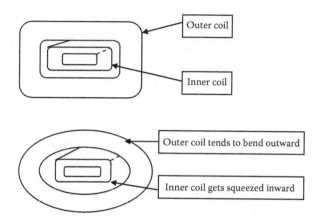

FIGURE 15.15
Forces on concentrically wound coils [1,2,4].

where F = crest outward force on inside conductor, in pound per inch of length of conductor; and l = the axial length of the primary coil in inches.

15.3.5 Mechanical Strength of Copper

Copper used in manufacturing windings for transformers is generally soft material with indefinite elastic limit. It work hardens easily. The modulus of elasticity for such materials is called the *tangent modulus of elasticity*. These material properties become worse with an increase in temperature. In practice, large forces exist only during short-circuit conditions. These forces are limited by protective circuits to less than 1 sec. Also, the change in the value of force takes in 1/240th of a second. Therefore, the first time a short-circuit load that exceeds the elastic limit of the copper is applied to the coils of a transformer, the coils deflect and take permanent shape in the first cycle. The maximum stress possible is 10,000 psi (Figure 15.15).

15.3.6 Backup Strength of Outer Turns

In a transformer configured in a core-type form, the outside coil is backed up by the outside turns. Also, leakage flux density diminishes for each turn outside of the inside turn and reduces to zero at the outermost turn. The average force of the whole body of turns is only half that on the inside turns and can have lower stress values for designs with higher voltages.

15.3.7 Compression Force on Inner Coil

This force is equal and opposite to the outward force on the outer coil. This force will push the inner coil onto the core. It is imperative to note that additional internal support can be easily provided to such a coil.

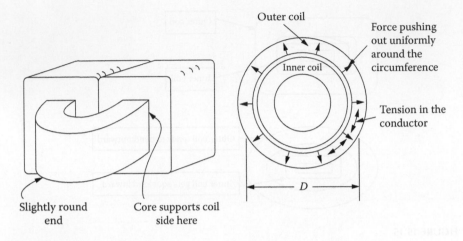

FIGURE 15.16
Inner coil subjected to compression, outer coil subjected to radial expansion [1,2,4].

15.3.8 Axial Displacement of Coils and Resultant Axial Forces

Under ideal conditions when both coils are of equal lengths, no net axial force can exist. However, under practical conditions, although every effort is made by a designer to match the axial lengths, if a distorted structure exists as shown in Figure 15.16, a net axial force can exist. In addition, tapping windings can cause a shift in the magnetic centers of coils which results in axial force (Figure 15.17).

15.3.9 Calculation of Axial Forces

1. Calculate first the total outward or radial force on the coil by taking half the force on the inner conductor and multiplying that by the total number of conductors.
2. Assume coils to be displaced as in Figure 15.23 in the vertical direction. The radial force can then be assumed to have the vertical component as shown and can be calculated as

$$F_{axial} = \text{Total radial force} \times \text{Sine of the angle } \Phi \qquad (15.17)$$

15.3.10 Short-Circuit Currents and Short-Circuit Capability

The ASA standard C 57.12-08.400 and any revisions thereafter indicate that the short-circuit force cannot be more than 25 times the *base current*. This base current is not the rated current of the transformer but the current rating of the self-cooled rating of the transformer. This condition is especially important for autotransformers and regulating transformers.

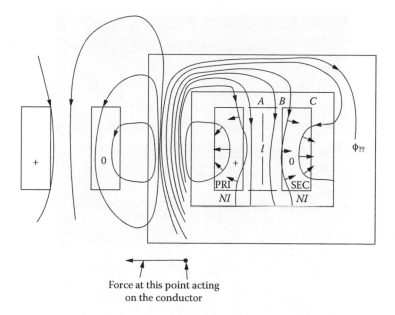

Force at this point acting
on the conductor

FIGURE 15.17
Forces exerted by magnetic flux lines [1,2,4].

In general, the 25 times condition evaluates to 18.75 times the rating at the forced-air cooling and 15 times the rating with forced oil, forced air cooling.

The above calculations also assume that the short circuit is applied at voltage zero, so that current is initially fully offset and will reach its crest value. It is less than twice the normal if the transformer resistance is large compared to the reactance. Also, under Y-Δ connections of the transformer windings, the line to ground short circuit gives rise to unusually large short-circuit currents. These forces are relieved by the pre-drying of spacers, and the insulation structure, as well as by subjecting the coils to pre-compression under clamps or presses (Figure 15.18).

15.4 Core Design

Typically for such solar energy alternative energy source applications, the cores are built of circular shapes as well as rectangular shapes as shown in Figure 15.19:

1. Circular wound cores
2. Toroidal-shaped cores
3. Rectangular-shaped cores

FIGURE 15.18
(See color insert) Coil clamping machine and fixture.

with CRGO Si steel or hot-rolled (HR) steel depending on the flux densities required. The former is used for flux densities close to 1.7 tesla (T), while the latter is used when the flux density requirements are lower, say up to 1.3 to 1.5 T. The CRGO Si steel saturates at about 15% more than the nominal value and HR steel at about 10% than the nominal value.

Figure 15.20 shows DPV-GT core cross sections for a three-limb core construction, and Table 15.8 presents the corresponding step widths.

In some cases amorphous core material is used by certain manufacturers up to about 1000 KVA, 33 kV. Many US utilities generally will not accept such designs because the core is brittle and it may cause ferro-resonance and that will in turn introduce additional harmonic content in the system besides that contributed by inverters in the system. In such cores the nonlinear inductance value is much smaller, and the related line capacitance creates a resonant frequency very close to 60 or 50 Hz and its related harmonics may

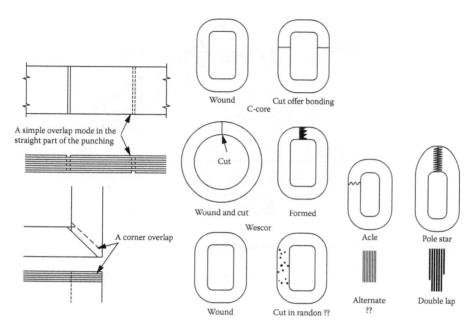

FIGURE 15.19
Different shapes of cores for DPV-GTs.

be up to the hundredth such component. These calculations can be verified by using the equation

$$\omega = 1 / \left(2\pi \operatorname{sqrt}(LC) \right) \qquad (15.18)$$

The additional concern for cores of DPV-GTs is the related seismological conditions in many areas. The cores have to withstand forces as high as 2 gs. The winding and core can be designed to withstand such forces with adequate winding anchoring and core clamping schemes.

Various factors must be evaluated when choosing a transformer so that it meets the needs of both the load and the application.

The time spent in selecting a transformer seems to be in direct proportion to the size of the unit. All too often, small transformers are selected with just a cursory look at the connected loads, and frequently a decision is made to choose one with the next larger kVA rating than the anticipated load. Conversely, large transformers, such as those used in electric utility applications, are closely evaluated because they represent large investments.

Most transformers used in commercial and industrial facilities fall in the middle ground, and they usually have ratings between 250 and 1000 kVA. On larger projects, they can go up to 10 MVA. Because these transformers represent the majority, you should evaluate them carefully before choosing a unit for a specific project and/or application.

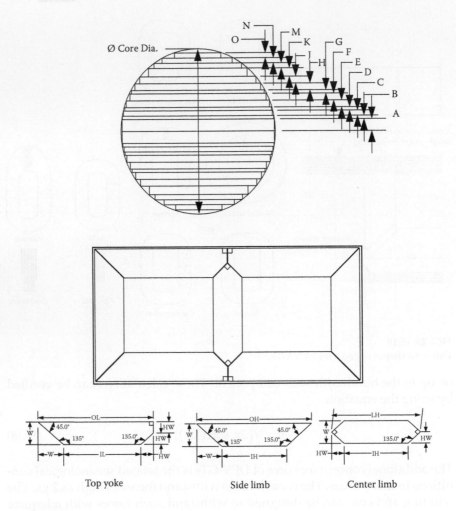

FIGURE 15.20
Core sections for three-phase industrial transformers, showing cross sections with different lamination widths and stack thicknesses for each width, and plan layout of laminations.

15.4.1 Selection Process

There are three main parameters in choosing a transformer:

- That it has enough capacity to handle the expected loads (as well as a certain amount of overload)
- That consideration be given to possibly increasing the capacity to handle potential load growth
- That the funds allocated for its purchase be based on a certain life expectancy (with consideration to an optimal decision on initial, operational, and installation costs)

TABLE 15.8

(a) DPV-GT Core Section Stack Thickness Details for Figure 15.20

Reference Number	Core Diameter (cm)	Total Stack Thk. 'Z' (mm)	Net Area with 0.97 Space Factor (cm²)		Stack Width (mm)												
					A	B	C	D	E	F	G	H	K	M	N	O
1	28	252	555.4	LEG	270	260	250	240	220	200	180	150	120			
				YOKE	270	260	250	240	220	200	180	150	150			
2	30	274	614.6	LEG	290	280	260	250	230	220	200	180	150	120		
				YOKE	290	280	260	250	230	220	200	180	150	150		
3	32	296	735.6	LEG	310	300	290	270	250	230	210	180	150	120		
				YOKE	310	300	290	270	250	230	210	180	150	150		
4	34	314	829.5	LEG	330	320	310	290	270	250	220	200	160	130		
				YOKE	330	320	310	290	270	250	220	200	160	160		
5	36	330	931.0	LEG	350	340	330	320	290	280	260	240	210	180	140	
				YOKE	350	340	330	320	290	280	260	240	210	180	180	
6	38	352	1040.8	LEG	370	350	340	320	320	280	260	240	210	180	140	
				YOKE	370	350	340	320	320	280	260	240	210	180	180	
7	40	370	1153.3	LEG	390	380	370	350	350	300	280	250	220	190	150	
				YOKE	390	380	370	350	350	300	280	250	20	190	190	
8	42	384	1268.4	LEG	410	400	390	380	380	340	320	300	270	250	220	170
				YOKE	410	400	390	380	380	340	320	300	270	250	220	220
9	44	390	1371.2	LEG	430	420	410	390	390	350	330	300	270	240	200	220
				YOKE	430	420	410	390	390	350	330	300	270	240	200	200

Continued

TABLE 15.8 (Continued)

(b) DPV-GT Core Section Stack Thickness Details for Figure 15.20

Reference Number	Core Diameter (cm)	Total Stack Thk. 'Z' (mm)	Net Area with 0.97 Space Factor (cm²)		A	B	C	D	E	F	G	H	K	M	N	O	Flitch Plate Size (mm)
1	28	252	555.4	LEG	38	14	12	9	14	11	10	11	7				75 × 6
				YOKE	38	14	12	9	14	11	10	11	7				
2	30	274	614.6	LEG	39	15	21	8	14	5	10	8	10	7			75 × 6
				YOKE	39	15	21	8	14	5	10	8	10	7			
3	32	296	735.6	LEG	40	16	12	18	14	12	9	12	9	6			75 × 6
				YOKE	40	16	12	18	14	12	9	12	9	6			
4	34	314	829.5	LEG	41	17	12	19	15	12	14	8	12	7			75 × 6
				YOKE	41	17	12	19	15	12	14	8	12	7			
5	36	330	931.0	LEG	43	17	12	11	17	14	11	10	12	9	9		100 × 6
				YOKE	43	17	12	11	17	14	11	10	12	9	9		
6	38	352	1040.8	LEG	44	30	11	18	14	12	10	9	11	9	8		100 × 6
				YOKE	44	30	11	18	14	12	10	9	11	9	8		
7	40	370	1153.3	LEG	45	18	13	21	17	19	10	14	11	8	9		100 × 6
				YOKE	45	18	13	21	17	19	10	14	11	8	9		
8	42	384	1268.4	LEG	46	19	13	12	19	15	13	10	14	8	10	13	125 × 6
				YOKE	46	19	13	12	19	15	13	10	14	8	10	13	
9	44	390	1371.2	LEG	47	19	14	22	18	14	12	15	13	11	10		140 × 6
				YOKE	47	19	14	22	18	14	12	15	13	11	10		

Stack Thickness (mm) columns: A, B, C, D, E, F, G, H, K, M, N, O

(c) DPV-GT Core Section Stack Thickness Details for Figure 15.20

Number	Core Circle Diameter (cm)	Stack Thickness (cm)	Sp. Factor	Net Area (cm²) Leg	Net Area (cm²) Yoke	Number of Steps Leg/Yoke
1	9.9	8.6	0.97	65.8	75.2	5/2
2	10.8	9.0	0.97	77.0	87.3	5/2
3	11.7	10.6	0.97	94.9	99.1	6/4
4	13.3	12.4	0.97	122.6	129.6	6/4
5	14.3	13.4	0.97	143.0	149.2	7/5
6	15.2	14	0.97	160.8	168.2	7/5
7	16.4	15.2	0.97	189.5	195.7	7/5
8	17.6	16.2	0.97	220.0	226.0	8/6
9	18.5	17.2	0.97	241.1	247.7	8/6
10	19.7	18	0.97	270.6	281.5	8/6
11	20.7	19	0.97	304.4	313.5	8/6
12	21.7[a]	20.2	0.97	333.7	339.9	8/6
13	23.8	22	0.95	390.4	404.1	8/6
14	25.2	23.2	0.95	439.3	454.7	8/6
15	28.0	25.2	0.97	555.4	555.4	9/9
16	30.0	27.2	0.97	641.6	641.6	10/10
17	32.0	29.6	0.97	735.6	735.6	10/10
18	34.0	31.4	0.97	829.5	829.5	10/10
19	36.0	33.0	0.97	931.0	931.0	11/11
20	38.0	35.2	0.97	1040.8	1040.8	11/11
21	40.0	37.0	0.97	1153.3	1153.3	11/11
22	42.0	38.4	0.97	1371.2	1371.2	12/12
23	44	39	0.97	1371.2	1371.2	11/11

[a] Core size used in design.

Both capacity and cost relate to a number of factors that should be evaluated. These include

- Application of the unit: where exactly is the DPV-GT being used?
- Choice of insulation type (liquid-filled or dry type): generally, dry-type transformers are preferred for mining applications
- Choice of winding material (copper or aluminum): for lower-rating transformers aluminum windings are preferred to lower cost
- Possible use of low-loss core material: CRGO or in some cases amorphous core material can be used for DPV-GTs
- Regulation (voltage stability): this is generally less than 5%
- Life expectancy: this is typically close to twenty years for DPV-GTs
- Any overloading requirements
- Basic insulation level (BIL): Class I as per IEEE standards
- Temperature considerations: 55/65°C for most outdoor applications
- Losses (both no-load and operating losses): to match the guarantee as specified by the end user
- Any nonlinear load demand: more experienced by DPV-GTs
- Shielding: may be typically used for 66 kV windings to even out voltage stress to achieve a better alpha value
- Accessories

15.4.2 Application of the Unit

The type of load and the transformer's placement are two key considerations that must be understood. For example, if the unit will be used for heavy welding service, such as in an automotive plant, very rigid construction will be called for because the coils will experience very frequent short-circuit-type loads; thus, good short-term overload capability may be required.

Sizing a transformer for a particular application with regard to the unit's life expectancy requires a good understanding of its insulation characteristics and the winding temperature due to loading. This, in turn, requires a careful analysis of the load profile (covering amplitude, duration, and the extent of linear and nonlinear loads).

The standard parameters for transformers operating under normal conditions include

- Nominal values of input voltage and frequency
- Approximately sinusoidal input voltage
- Load current with a harmonic factor not exceeding 0.05 p.u.
- Installation at an altitude of less than 1000 m (3300 ft)

- No damaging fumes, dust, vapor, etc., in installed environment specified by the IP level
- An ambient temperature that does not exceed 30°C as a daily average or 40°C at any time, and which does not fall below –20°C
- Overloads within acceptable levels of ANSI/IEEE loading guidelines (dry or liquid)

If some of the above conditions cannot be met in a particular application, then you should work closely with the manufacturer so that the selected transformer's operating characteristics and/or size will compensate for the particular situation. For example, if the ambient temperature will exceed standard conditions or if the unit will be installed at a high elevation, then an appropriate solution might be to specify a transformer that is rated higher than what the load requires, in effect underutilizing the unit to compensate for the local conditions.

15.4.3 Choice of Liquid-Filled or Dry Type

Information on the pros and cons of the available types of transformers frequently varies depending upon which manufacturer you are talking to and what literature you are reading. Nevertheless, there are certain performance and application characteristics that are almost universally accepted.

Basically, there are two distinct types of transformers: liquid insulated and cooled (liquid-filled type) and nonliquid insulated, air or air/gas cooled (dry type). Also, there are subcategories of each main type.

For liquid-filled transformers, the cooling medium can be conventional mineral oil. There are also wet-type transformers using less flammable liquids, such as high fire point hydrocarbons and silicones.

Liquid-filled transformers are normally more efficient than dry types, and they usually have a longer life expectancy. Also, liquid is a more efficient cooling medium in reducing hot-spot temperatures in the coils. In addition, liquid-filled units have a better overload capability.

There are some drawbacks, however. For example, fire prevention is more important with liquid-type units because of the use of a liquid cooling medium that may catch fire. (Dry-type transformers can catch fire, too.) It is even possible for an improperly protected wet-type transformer to explode. And, depending on the application, liquid-filled transformers may require a containment trough for protection against possible leaks of the fluid.

Because of the above reasons, and because of the ratings, indoor-installed distribution transformers of 600 V and below usually are dry types.

Arguably, when choosing transformers, the changeover point between dry types and wet types is between 500 kVA and about 2.5 MVA, with dry types used for the lower ratings and wet types for the higher ratings. Important factors when choosing what type to use include where the transformer will be installed, such as inside an office building or outside, servicing an industrial load. Dry-type transformers with ratings exceeding 5 MVA are available, but the vast majority of the higher-capacity transformers are liquid filled. For outdoor applications, wet-type transformers are the predominant choice.

Dry-type transformers come in enclosures that have louvers or are sealed. Here, subcategories include different methods of insulation such as conventional varnish, vacuum pressure impregnated (VPI) varnish, epoxy resin, or cast resin insulation systems.

The insulation system for liquid-filled distribution transformers is typically composed of enameled wire, cellulose paper impregnated with a dielectric liquid, and the liquid itself. The dielectric-grade paper most often used is derived from sulfate (kraft) wood pulp from softwoods. With the introduction of dicydianamid to the paper-making process, the standard temperature winding rise is now 65°C.

The ambient temperature base in the United States is a 30°C average over a 24-h period with a 40°C maximum. The present allowable hot-spot temperature (the difference between the average winding temperature rise and the hottest spot in the windings) is 15°C. Thus, the permitted operating hot-spot temperature, based on an average ambient temperature of 30°C, is 110°C.

New synthetic insulating materials are leading to even higher permitted hot spots. These materials include polyester, fiberglass, and more commonly, aramid paper. *Aramid paper* is a term applied generically for wholly aromatic polyamide paper. To keep costs reasonable while still achieving gains in acceptable hot-spot temperature limits, both aramid and thermally upgraded kraft paper are used together in a hybrid insulation system. As of this writing, new liquid-filled transformers, called high-temperature transformers (HTTs) are being built using this technology. The temperature rise of HTTs is an average winding rise of 115°C over a 30°C average ambient. Factoring in the temperature difference (20°C) between the average winding temperature (145°C) and the hot-spot temperature, the maximum temperature (165°C) will be at a level that is higher than the fire point of conventional transformer oil (mineral oil). For this reason, it is recommended that fire-resistant fluids be used for HTTs.

The process of properly impregnating the paper with a liquid is a standard manufacturing operation. The core/coil assembly is mounted in the tank, lead assemblies are attached, and the filling process begins. A partial vacuum is produced while the secondary leads circulate current to heat the coils and drive out any excess moisture. Later, while still under vacuum, heated degasified and filtered dielectric liquid is introduced. After filling and additional vacuum time, the tank cover is sealed in place. The headspace between the liquid surface and the tank cover, which allows for expansion and contraction due to thermal cycling, can be specified to be dry nitrogen gas in larger units.

15.4.4 Environmental Concerns

For liquid-filled transformers containing more than 660 gal, the Environmental Protection Agency (EPA) requires some type of containment be used to control possible leaks of the liquid. Environmentally unfriendly fluids, such as polychlorinated biphenyls (PCBs) and chlorofluorocarbons (CFCs), have been banned or are severely restricted, replaced for the most part by nontoxic, nonbioaccumulating, and nonozone-depleting fluids, such as fire-resistant silicones and fire-resistant hydrocarbons. These fluids are not covered by the Resource Conservation and Recovery Act (RCRA); however, they are covered by the Clean Water Act (CWA).

Some transformer liquids (known as nonflammable-type fluids) are covered under both the RCRA and the CWA, and certain requirements may be called for regarding special handling, spill reporting, disposal procedures, and record keeping. These fluids also require provisions for special transformer venting. As such, the above factors can have an effect on installation costs, long-term operating costs, and maintenance procedures.

15.4.5 Liquid Dielectric Selection Factors

The selection of which liquid dielectric coolant to use is driven primarily by economics and codes. Conventional mineral oil is most often specified as it is very economical and, unless subject to unusual service, maintains acceptable performance for decades.

As it is possible that a high-energy arc can occur in a transformer, fire safety becomes an important issue. When conventional mineral oil is restricted (usually due to fire code requirements), less-flammable fluids often are used. The most popular are fire-resistant hydrocarbons (also known as high-molecular-weight hydrocarbons) and 50 cSt (a viscosity measurement unit) silicone fluids. Other fluids include high fire point polyol esters and poly-alpha olefins. In addition to safety considerations, you should evaluate performance factors for liquid-filled transformers in regard to dielectric strength and heat transfer capabilities of the fluid. The fire-resistant hydrocarbon fluids have been widely used in power-class transformations though 60 MVA and have over a 500 kV BIL.

At one time, askarel fluid, a generic term for a group of certain fire-resistant electrical insulating liquids, including often used PCBs, represented the standard for fire safety in liquid dielectrics. But, PCBs were banned because of toxicity and environmental concerns.

15.4.6 Cast Coil Insulation Systems

Dry-type transformers can have their windings insulated various ways. A basic method is to preheat the conductor coils and then, when heated, dip them in varnish at an elevated temperature. The coils are then baked to cure the varnish. This process is an open-wound method and helps ensure penetration of the varnish. Cooling ducts in the windings provide an efficient and economical way to remove the heat produced by the electrical losses of the transformer by allowing air to flow through the duct openings. This dry-type insulation system operates satisfactorily in most ambient conditions found in commercial buildings and many industrial facilities.

When greater mechanical strength of the windings and increased resistance to corona (electrical discharges caused by the field intensity exceeding the dielectric strength of the insulation) are called for, VPI of the varnish forces the insulation (varnish) into the coils by using both vacuum and pressure. Sometimes, for additional protection against the environment (when the ambient air can be somewhat harmful), the end coils are also sealed with an epoxy resin mixture.

A cast coil insulation system, another version of the dry-type transformer, is used when additional coil strength and protection are advisable. This type of insulation is used for transformers located in harsh environments such as cement and chemical plants and outdoor installations where moisture, salt spray, corrosive fumes, dust, and metal particles can destroy other types of dry-type transformers. These cast coil units are better able to withstand heavy power surges, such as frequent but brief overloads experienced by transformers serving transit systems and various industrial machinery. Cast coil units also are being used where previously only liquid-filled units were available for harsh environments. They can have the same high levels of BIL while still providing ample protection of the coils and the leads going to the terminals.

Unlike open-wound or VPI transformers, cast coil units have their windings completely cast in solid epoxy. The coils are placed into molds and cast, usually under vacuum. The epoxy is a special type that keeps the coils protected from corrosive atmospheres and moisture, as well as keeps the coils secure from the high mechanical forces associated with power surges and short circuits. Mineral fillers and glass fibers are added to the pure epoxy to give it greater strength. Flexibilizers are also added to improve its ability to expand and contract with the coil conductors for proper operation of the transformer under various load conditions.

Different manufacturers use different epoxy filling materials and in different amounts. Important factors that manufacturers must consider when choosing filler material and the proportion to use include

- Temperature rating of the transformer
- Mechanical strength of the coils
- Dielectric strength of the insulation
- Expansion rate of the conductors under various loadings
- Resistance to thermal shock of the insulation system

Cast coil transformers consist of separately wound and cast high- and low-voltage coils. During manufacture, the high-voltage coil winding wires are placed in a certain pattern using preinsulated wire. The completely wound coil is then placed in a mold designed to form a heavy coating of epoxy around the coil. After vacuum filling of the epoxy, the mold is placed in an oven for a number of hours to allow the epoxy to cure and achieve full hardness and strength.

There are two types of low-voltage windings available, both of which provide protection from hostile environments. One type is vacuum cast like the high-voltage winding. The other type uses a *nonvacuum* technique of epoxy application to achieve strength. Sheet insulation, such as Nomex® or fiberglass, is impregnated with uncured epoxy and then interleaved on the heavy, low-voltage conductors to literally "wind-in" the epoxy. During oven curing of the low-voltage coil, the epoxy flows onto the conductor and cures into a solid cylinder of great strength. These nonvacuum coils are then fully sealed by pouring epoxy into the margins or ends of the windings. Both procedures provide good protection from hostile environments.

Because of the additional materials and procedures associated with manufacturing cast coil transformers, they cost more. However, cast coil transformers are designed to operate with lower losses, require less maintenance than regular types, and effectively operate in environments that may cause early failure with conventional dry types. Also, cast coil transformers, if operated properly, will normally have a longer life expectancy than other dry-type transformers.

15.4.7 Choice of Winding Material

A transformer's coils can be wound with either copper or aluminum conductors. For equivalent electrical and mechanical performance, aluminum-wound transformers usually cost less than copper-wound units. Because copper is a better conductor, a copper-wound transformer can be at times slightly smaller than its aluminum counterpart, for transformers with equivalent electrical ratings, because the copper conductor windings will be smaller. However, most manufacturers supply aluminum and copper transformers in the same enclosure size.

Aluminum-wound transformers are by far the majority choice in the United States. With both materials, the winding process and the application of insulation are the same. Connections to the terminals are welded or brazed. Coils made of copper wires have slightly higher mechanical strength.

You should determine the transformer manufacturer's experience in building its products and if the firm has a proven record in using both types of conductors. This is especially true of manufacturers of dry-type units.

15.4.8 Use of Low-Loss Core Material

Choice of metal is critical for transformer cores, and it is important that good-quality magnetic steel be used. There are many grades of steel that can be used for a transformer core. Each grade has an effect on efficiency on a per-pound basis. The choice depends on how you evaluate nonload losses and total owning costs.

Almost all transformer manufacturers today use steel in their cores which provides low losses due to the effects of magnetic hysteresis and eddy currents. To achieve these objectives, high permeability, cold-rolled, grain-oriented, silicon steel is almost always used. Construction of the core utilizes step lap mitered joints, and the laminations are carefully stacked.

15.4.9 Amorphous Cores

A new type of liquid-filled transformer introduced commercially in 1986 uses ultra-low-loss cores made from amorphous metal; the core losses are between 60 and 70% lower than those for transformers using silicon steel. To date, these transformers have been designed for distribution operation primarily by electric utilities and use wound-cut cores of amorphous metal. Their ratings range from 10 kVA through 2500 kVA. The reason utilities purchase them, even though they are more expensive than silicon steel core transformers, is because of their high efficiency. U.S. utilities placed more than 400,000 amorphous steel core transformers in operation through 1995. The use of amorphous core liquid-filled transformers is now being expanded for use in power applications for industrial and commercial installations. This is especially true in other countries such as Japan.

Amorphous metal is a new class of material having no crystalline formation. Conventional metals possess crystalline structures in which the atoms form an orderly, repeated, three-dimensional array. Amorphous metals are characterized by a random arrangement of their atoms (because the atomic structure resembles that of glass, the material is sometimes referred to as glassy metal). This atomic structure, along with the difference in the composition and thickness of the metal, accounts for the very low hysteresis and eddy current losses in the new material.

Cost and manufacturing technique are the major obstacles for bringing to the market a broad assortment of amorphous core transformers. The price of these units is typically 15% to 40% higher than that of silicon steel core transformers. To a degree, the price differential is dependent upon which grade of silicon steel the comparison is being made. (The more energy efficient the grade of steel used in the transformer core, the higher the price of the steel.)

At present, amorphous cores are not being applied in dry-type transformers. However, there is continuous developmental work being done on amorphous core transformers, and the use of this special metal in dry-type transformers may become a practical reality sometime in the future.

If considering the use of an amorphous core transformer, the economic trade-off should be determined; in other words, the price of the unit versus the cost of losses. Losses are especially important when transformers are lightly loaded, such as during the hours from about 9 PM to 6 AM. When lightly loaded, the core loss becomes the largest component of a transformer's total losses. Thus, the cost of electric power at the location where such a transformer is contemplated is a very important factor in carrying out the economic analyses.

Different manufacturers have different capabilities for producing amorphous cores, and recently, some have made substantial advances in making these cores for transformers. The technical difficulties of constructing a core using amorphous steel have restricted the size of transformers using this material. The metal is not easily workable, being very hard and difficult to cut, thin and flimsy, and difficult to obtain in large sheets. However, development of these types of transformers continues; you can expect units larger than 2500 kVA to be made in the future.

15.4.10 Protection from Harsh Conditions

For harsh environments, whether indoor or outdoor, it is critical that a transformer's core/coil, leads, and accessories be adequately protected.

In the United States, almost all liquid-filled transformers are of sealed-type construction, automatically providing protection for the internal components. External connections can be made with *dead front* connectors that shield the leads. For highly corrosive conditions, stainless steel tanks can be employed.

Dry-type transformers are available for either indoor or outdoor installation. Cooling ducts in the windings allow heat to be dissipated into the air. Dry types can operate indoors under almost all ambient conditions found in commercial buildings and light manufacturing facilities.

For outdoor operations, a dry-type transformer's enclosure will usually have louvers for ventilation. But, these transformers can be affected by hostile environments (dirt, moisture, corrosive fumes, conductive dust, etc.) because the windings are exposed to the air. However, a dry-type transformer can be built using a sealed tank to provide protection from harmful environments. These units operate in their own atmosphere of nonflammable dielectric gas.

Other approaches to building dry-type transformers for harsh environments include cast coil units, cast resin units, and vacuum pressure encapsulated (VPE) units, sometimes using a silicone varnish. Unless the dry-type units are completely sealed, the core/coil and lead assemblies should be periodically cleaned, even in nonharsh environments, to prevent dust and other contaminant buildup over time.

15.4.11 Insulators

Dry-type transformers normally use insulators made from fiberglass-reinforced polyester molding compounds. These insulators are available up to a rating of 15 kV and are intended to be used indoors or within a moisture-proof enclosure. Liquid-filled transformers employ insulators made of porcelain. These are available in voltage ratings exceeding 500 kV. Porcelain insulators are track resistant, suitable for outdoor use, and easy to clean.

High-voltage porcelain insulators contain oil-impregnated paper insulation that acts as a capacitive voltage divider to provide uniform voltage gradients. Power factor tests must be performed at specific intervals to verify the condition of these insulators.

15.4.12 Regulation

The difference between the secondary's no-load voltage and full-load voltage is a measure of the transformer's regulation. This can be determined by using the following equation:

$$\text{Regulation}(\%) = (100)\big([V.sub.nl] - [V.sub.fl]\big)\big/\big([V.sub.fl]\big) \qquad (15.19)$$

when *V.sub.fl* is maintained constant, and where [*V.sub.nl*] is the no-load voltage and [*V.sub.fl*] is the full-load voltage. Poor regulation means that as the load increases, the voltage at the secondary terminals drops substantially. This voltage drop is due to resistance in the windings and leakage reactance between the windings. However, good regulation may offer some other problems.

Voltage regulation and efficiency are improved with low impedance, but the potential for serious damage also goes up. Sometimes manufacturers, in order to meet demands for good regulation, design transformers with leakage reactance as low as 2%. A transformer so designed is liable to be severely damaged if a short circuit occurs on the transformer's secondary, especially if the total power on the system is large (a stiff source with low impedance).

The mechanical stresses in a transformer vary approximately as the square of the current. Stresses in a transformer resulting from a short circuit could be approximately six times as great in a transformer having 2% impedance as they would be in one having 5% impedance (where reactance is the major component of the impedance voltage drop).

Of course, a good circuit protection scheme can address this problem. Short-circuit integrity is readily available if you wish to include in your transformer specifications that it follow the ANSI/IEEE Guide for Short Circuit Testing; C57.12.90-1993 for wet units and C57.12.91-1995 for dry units.

15.4.13 Voltage Taps

Even with good regulation, the secondary voltage of a transformer can change if the incoming voltage changes. Transformers, when connected to a utility system, are dependent upon utility voltage; when utility operations change or new loads are connected to their lines, the incoming voltage to your facility may decrease, or even perhaps increase.

To compensate for such voltage changes, transformers are often built with load tap changers (LTCs), or sometimes, no-load tap changers (NLTCs).

(LTCs operate with the load connected, whereas NLTCs must have the load disconnected.) These devices consist of taps or leads connected to either the primary or secondary coils at different locations to supply a constant voltage from the secondary coils to the load under varying conditions.

Tap changers connected to the primary coils change the connections from the incoming line to various leads going to the coils. When tap changers are connected to the secondary coils, the changing of connections is made from the coils to the output conductors.

Tap changers can be operated by either manual switching or by automatic means. Transformers with tap changers usually have a tap position indicator to allow you to know what taps are being used.

15.4.14 Life Expectancy

There's a common presumption that the useful life of a transformer is the useful life of the insulating system, and that the life of the insulation is related to the temperature being experienced. You should recognize that the temperatures of the windings vary; there are hot spots usually at an accepted maximum 30°C above average coil winding temperature for dry-type transformers. The hot-spot temperature is the sum of the maximum ambient temperature, the average winding temperature rise (where the winding refers to the conductor), and the winding gradient temperature (the gradient being the differential between the average winding temperature rise and the highest temperature of the winding).

The nameplate kVA rating of a transformer represents the amount of kVA loading that will result in the rated temperature rise when the unit is operated under normal service conditions. When operating under these conditions (including the accepted hot-spot temperature with the correct class of insulation materials), you should achieve a normal life expectancy for the transformer.

Information on dry-type transformer loading from ANSI/IEEE C57.96-1989 indicates that you can have a twenty-year life expectancy for the insulation system in a transformer. However, due to degradation of the insulation, a transformer might fail before twenty years have elapsed. For dry-type transformers having a 220°C insulating system and a winding hot-spot temperature of 220°C, and with no unusual operating conditions present, the twenty-year life expectancy is a reasonable time frame. [The 220°C represents a transformer used in a location with a 40°C (104°F) maximum ambient temperature, an average 150°C rise in the conductor windings, and a 30°C gradient temperature.]

Most 150°C rise dry-type transformers are built with 220°C insulation systems. Operating such a transformer at rated kVA on a continuous basis with a 30°C average ambient should equate to a normal useful life. (Note that 40°C maximum ambient in any 24-hour period with 30°C as the 24-hour average is considered a standard ambient.)

When based solely on thermal factors, the life of a transformer increases appreciably if the operating temperature is lower than the maximum temperature rating of the insulation. However, you should recognize that the life expectancy of transformers operating at varying temperatures is not accurately known. Fluctuating load conditions and changes in ambient temperature make it difficult, if not impossible, to arrive at such definitive information.

15.4.15 Overloading

For effective operation of an electrical system, transformers are sometimes overloaded to meet operating conditions. As such, it is important that you have an understanding with the transformer manufacturer as to what overloading the unit can withstand without causing problems.

The main problem is heat dissipation. If a transformer is overloaded by a certain factor, say 20% beyond kVA rating for a certain period of time, depending upon that period of time, it is probable that any heat developed in the coils will be transferred easily to the outside of the transformer tank. Therefore, there is a reasonable chance that the overloading will not cause a problem. However, when longer time periods are involved, heat will start to build up internally within the transformer, possibly causing serious problems.

An effective way of removing this heat is to use built-in fans; this way, the load capability can be increased without increasing the kVA rating of the transformer.

Dry-type transformers typically have a fan-cooled rating that is 1.33 times the self-cooled rating. Some transformer designs can provide ratings of 1.4 to 1.5 times self-cooled units. If you have such requirements, you should prepare a carefully written specification.

Liquid-filled transformers, because of their double heat-transfer requirement (core/coil-to-liquid and liquid-to-air), have a lower forced air rating. Usually, the increased rating is 1.15 times the self-cooled rating for small units and 1.25 times self-cooled rating for larger "small power transformers." When above 10 MVA, the ratio may be as high as 1.67 to 1.

You should recognize two distinct factors when forced cooling is used. First, the concept is used to obtain a higher transformer capacity, but when doing so, losses are increased substantially. A dry-type transformer operating at 133% of its self-cooled rating will have conductor losses of nearly 1.8 times the losses at the self-cooled rating. And, there will be some losses in the form of power to operate the fan motors. The normal no-load losses remain constant regardless of the load. The other liability is that when additional equipment is used, such as fans, the chance of something malfunctioning increases.

Table 1, on page 50, lists the loading capability for liquid-filled, 65°C rise transformers, based on normal loss of life. This information is from Table 5 in ANSI/IEEE C57.91-1981, Guide for Loading Mineral Oil Immersed Power Transformers Rated 500 kVA and Less.

Table 2, on page 50, in ANSI/IEEE C57.91-1981 lists the loading capability for 200°C dry-type insulation system transformers, based on normal loss of life. This information is from Table 6 in ANSI/IEEE C57.96-1989, Guide for Loading Dry-Type Transformers.

15.4.16 Insulation Level

The insulation level of a transformer is based on its basic impulse level (BIL). The BIL can vary for a given system voltage, depending upon the amount of exposure to system over-voltages, a transformer might be expected to encounter over its life cycle. ANSI/IEEE Standards C57.12.00-1993 and C57.12.01-1989 indicate the BILs that may be specified for a given system voltage. You should base your selection on prior knowledge with similar systems, or on a system study such as that performed by qualified engineering firms or by selecting the highest BIL available for the system's voltage.

If the electrical system in question includes solid-state controls, you should approach the selection of BIL very carefully. These controls, which when operating chop the current, may cause voltage transients.

15.4.17 Liquid-Filled Temperature Considerations

Liquid-type transformers use insulation based on a cellulose/fluid system. The fluid serves as both an insulating and cooling medium. Forms are used (which are rectangular or cylindrical shaped) when constructing the windings, and spacers are used between layers of the windings. The spacing is necessary to allow the fluid to flow and cool the windings and the core.

For cooling, fluid flows in the transformer through ducts and around the coil ends within a sealed tank that encompasses the core and coils. Removal of the heat in the fluid takes place in external tubes, usually elliptical in design, welded to the outside tank walls.

When transformer ratings begin to exceed 5 MVA, additional heat transfer is required. Here, radiators are used; they consist of headers extending from the transformer tank on the bottom and top, with rows of tubes connected between the two headers. The transformer fluid, acting as a cooling medium, transfers the heat picked up from the core and coils and dissipates it to the air via the tubes.

The paper insulation used today in liquid-filled transformers is thermally upgraded, allowing a 65°C average winding temperature rise as standard. Until the 1960s, a 55°C rise was the standard.

Sometimes, transformer specifications are written for a 55/65°C rise. This provides an increase in the operating capacity by 12% since the kVA specified is based on the old 55°C rise basis but the paper supplied is the thermally upgraded kraft type.

For both wet- and dry-type units, a key factor in transformer design is the amount of temperature rise that the insulation can withstand. Lower temperature rise ratings of transformers can be achieved in two ways: by increasing the conductor size of the winding (which reduces the resistance and therefore the heating) or by derating a larger, higher temperature rise transformer. Be careful when using the latter method—because the percent impedance of a transformer is based on the higher rating, the let-through fault current and start-up inrush current will be proportionately higher than the rating at which it is being applied. Consequently, downstream equipment may need to have a higher withstand and interrupting rating, and the primary breaker may need to have a higher trip setting in order to hold in on start-up.

The lower temperature rise transformers are physically larger and, therefore, will require more floor space. On the plus side, a lower temperature rise transformer will have a longer life expectancy. The latest energy codes recommend selecting transformers to optimize the combination of no-load, part-load, and full-load losses without compromising the operational and reliability requirements of the electrical system.

15.4.18 Dry-Type Temperature Considerations

Dry-type transformers are available in three general classes of insulation. The main features of insulation are to provide dielectric strength and to be able to withstand certain thermal limits. Insulation classes are 220°C (Class H), 185°C (Class F), and 150°C (Class B). Temperature rise ratings are based on full-load rise over ambient (usually 40°C above ambient) and are 150°C (available only with Class H insulation), 115°C (available with Class H and Class F insulation), and 80°C (available with Class H, F, and B insulation). A 30°C winding hot-spot allowance is provided for each class.

The lower temperature rise transformers are more efficient, particularly at loadings of 50% and higher. Full load losses for 115°C transformers are about 30% less than those of 150°C transformers. And 80°C transformers have losses that are about 15% less than 115°C transformers and 40% less than 150°C transformers. Full load losses for 150°C transformers range from about 4% to 5% for 30 kVA and smaller to 2% for 500 kVA and larger.

When operated continuously at 65% or more of full load, the 115°C transformer will pay for itself over the 150°C transformer in two years or less (one year if operated at 90% of full load). The 80°C transformer requires operation at 75% or more of full load for a two-year payback, and at 100% load to payback in one year over the 150°C transformer. If operated continuously at 80% or more of full load, the 80°C transformer will have a payback over the 115°C transformer in two years or less (1.25 years at 100% loading).

A designer should note that at loadings below 50% of full load, there is essentially no payback for either the 115°C or the 80°C transformer over the 150°C transformer. Also, at loadings below 40%, the lower temperature rise transformers become less efficient than the 150°C transformers. Thus, not only is there no payback, but also the annual operating cost is higher.

15.4.19 Losses

Because the cost of owning a transformer involves both fixed and operating costs, and because the cost of electric power is constantly on the rise, the

TABLE 15.9

Performance Figures for DPV-GTs: Losses/Impedance

KVA	Material HV/LV	No Load Loss (W)	Load Loss	% Impedance
200	Al/Al	700	3680	4.75
250	Al/Al	700	4350	4.75
315	Al/Al	780	5700	4.75
315	Cu/Cu	810	5000	4.75
400	Al/Al	890	6600	4.75
400	Cu/Cu	970	5900	4.75
500	Al/Al	1080	7800	4.75
500	Cu/Cu	1220	6900	4.75

cost of energy lost over a period of time due to a transformer's losses can substantially exceed the purchase price of a unit. As such, it is important that transformer no-load and load losses be evaluated carefully (Table 15.9).

No-load losses consist of hysteresis and eddy currents in the core, copper loss due to no-load current in the primary winding, and dielectric loss. The core losses are the most important.

Load losses include FR loss in the windings, FR losses due to current supplying the losses, eddy current loss in the conductors due to leakage in the field, and stray losses in the transformer's structural steel. Specifying higher efficiency requires larger conductors for the coils to reduce FR losses. This means added cost, but the payback may be significant.

15.4.20 *k*-Factor

Some transformers (liquid and dry types) are now being offered with what is called a *k*-factor rating. This is a measure of the transformer's ability to withstand the heating effects of nonsinusoidal harmonic currents produced by much of today's electronic equipment and certain electrical equipment. Because of the problems created by harmonics, ANSI/IEEE, in the late 1980s, formulated the C57.110-1986 standard, Recommended Practice for Establishing Transformer Capability When Supplying Nonsinusoidal Load Currents. It applies to transformers up to 50 MVA maximum nameplate rating when these units are subject to nonsinusoidal load currents having a harmonic factor exceeding 0.05 per-unit (pu), the percent value of the base unit. (Harmonic factor is defined as the ratio of the effective value of all the harmonics to the effective value of the fundamental 60-Hz frequency.)

In December 1990, UL announced listings for dry-type general-purpose and power transformers affected by nonsinusoidal currents in accordance with the above ANSI/IEEE C57.110-1986. The "listing investigation" is directed to submitting transformers for testing to certain factors relating to rms current at certain harmonic orders in a specified way that correlates with heating losses. The factors involved in the tests are collectively called the k-factor.

Transformers meeting k-factor requirements also address the need for providing for high neutral currents. Because the neutral current may be considerably greater than the phase current, the neutral terminal of the transformer is sometimes doubled in size for additional customer neutral cables. It is important to recognize the impact caused by harmonic currents.

Oversized primary conductors are used to compensate for circulating harmonic currents. The secondary is also given special consideration. As the frequency increases to 180 Hz (as in the case with the third harmonic), and greater, the skin effect (where current begins to travel more on the circumference of the conductor) becomes more pronounced. To compensate for this, the windings are composed of several smaller sizes of conductor, with the circumference of the total conductors being greater. The transformer design also incorporates a reduction in core flux to compensate for harmonic voltage distortion.

For help in determining what k-factor to use when you specify a transformer while designing an electrical system for a facility, identify what harmonic-producing equipment is going into the system. Then, obtain information on the harmonic spectrum and the associated amplitudes produced by the offending apparatus from the manufacturer of the equipment.

Be careful when using k-rated transformers having abnormally low impedance, particularly those units with ratings of k-20 and higher. Such low impedance transformers can actually increase harmonics neutral current problems and even cause some loads to malfunction or cause damage to equipment. The use of abnormally low impedance transformers will act to significantly increase the neutral current and, therefore, negate some of the benefits of doubling the neutral conductors. It is important for isolation transformers to be used for high harmonic loads having normal (3% to 6%) impedance. Some highly knowledgeable engineers believe that it is wrong to specify transformers with ratings k-20 and higher for commercial office loads. If the harmonics are of such high magnitude and it is believed a transformer with a rating of k-20 or higher should be used, then careful attention should be given to make sure that the impedance of the unit is at least 3%.

15.4.21 Shielding

Depending upon the loads being served, the ability of a transformer to attenuate electrical noise and transients would be a helpful attribute. While what is commonly referred to as *dirty power* possibly cannot be stopped at the source causing the noise, corrective measures can be taken, including the application of a shield between the primary and secondary of a transformer. This type of construction is usually considered when a distribution transformer is serving solid-state devices such as computers and peripheral equipment.

There are two types of noise and voltage transients: common-mode noise and transients and normal or transverse-mode noise and transients. Common-mode power aberrations are disturbances between the primary lines and the ground (phase-to-ground); transverse-mode power aberrations are line-to-line disturbances. It is important to recognize this difference because an electrostatic shield will not reduce transverse disturbances. However, transverse disturbances are slightly reduced by a transformer's impedance, and this is true whether or not a transformer has a shield.

To substantially reduce transverse-mode power aberrations, surge suppressors are used to handle the transients, and filters are used to handle the noise. Some literature show voltage sine curves with disturbances imposed on the curve as well as clean voltage sine curves, and information is included to the effect that an electrostatic shield is responsible for reducing or eliminating the disturbances. This is incorrect because the voltage sine curve portrays line-to-line characteristics, and shielding has no effect upon such disturbances.

An electrostatic shield is a grounded metal barrier between the primary and secondary that filters common-mode noise, thus delivering cleaner power and reducing the spikes caused by common-mode voltage transients. The shield takes most of the energy from the voltage spike and delivers it to the ground. A number of authorities agree that transformers built to deliver a 60 dB (a 1000-to-1 ratio) reduction in common-mode disturbances (noise and voltage transients) will help solve or prevent such power aberrations from causing problems. Some transformers are built with the ability to provide a 100 dB (100,000-to-1 ratio) attenuation, and even larger ratios. If poor power quality may be a problem on a system where you plan to install a transformer, get information on the unit's attenuation ratio and verify that the power problem stems from common-mode disturbances.

An example of the effect of attenuation would be a lightning strike that induces a 1000 V spike on a power line connected to the primary of a transformer. A shield would take most of this energy to ground, and if the attenuation is 60 dB (1000-to-1 ratio), an approximate IV bump will be passed to the secondary and onto the feeder or branch circuit. A number of loads can take a bump of this magnitude without damage. If there is a branch circuit and another shielded transformer ahead of a load, the bump will be further reduced by the second transformer. This type of reduction is caused by an effect called *transformer cascading*.

15.4.22 Placing Transformers Near the Load

Locating a transformer indoors, on the rooftop, or adjacent to a building in order to minimize the distance between the unit and the principal load results in reducing energy loss and voltage reduction. It also reduces the cost of secondary cable. Such placements of high-voltage equipment require closer consideration of electrical and fire safety issues. These conflicting goals can be satisfied by using transformers permitted by code and insurance companies.

When liquid-filled transformers are preferred, less-flammable liquids are widely recognized for indoor and close building proximity installations. Wet-type transformers using less-flammable, or high fire point liquids, have been recognized by the NEC since 1978 for indoor installation without the need for vault protection unless the voltage exceeds 35 kV. Based on this type of transformer's excellent fire safety record, code and insurance restrictions have become minimal. Conventional mineral oil units are allowed indoors, but only if they are installed in a special 3-hour-rated vault (with a few exceptions) per the construction requirements of NEC Article 450, Part C. There is a requirement for liquid containment when wet-type transformers are used, regardless of the type of fluid employed.

When dry-type units are preferred, they have fewer code restrictions. Obviously, these types of transformers do not need liquid containment. Per the requirements listed in NEC Sec. 450-21, there are minimum clearances that you must observe, and units over 112.5 kVA require installation in a transformer room of fire-resistant construction, unless they are covered by one of two listed exceptions. As with liquid units, dry transformers exceeding 35 kV must also be located in a 3-hour-rated vault.

A liquid-filled transformer may experience leakage around gaskets and fittings; however, if the installation was carried out correctly, this should not be a problem. Major maintenance procedures may require inspection of internal components, meaning that the coolant will have to be drained. Coils in liquid-type units are much easier to repair than coils in dry-type transformers. Cast coils are not repairable; they must be replaced.

15.4.23 Accessories

Accessories are usually an added cost, and sometimes they are installed while the transformer is being built. Therefore, you should have some knowledge of accessories and incorporate in the transformer specification those accessories that, when installed, would be beneficial to the transformer's performance. Some of the accessories available include the following:

- Stainless steel tank and cabinet for extra corrosion protection (liquid-filled only)
- Special paint/finishes for corrosive atmospheres and ultraviolet light (liquid-filled only)
- Weather shields for outdoor units, protective provisions for humid environments, and rodent guards (dry-type only)
- Temperature monitors: There are a number of options available from simple thermometers to more extensive single- or three-phase temperature monitoring as well as options for contacts to initiate alarms and/or trip circuits as well as start cooling fans.

- Space heaters to prevent condensation during prolonged shutdown (usually with thermostats)
- Optional location of openings for primary and secondary leads
- Special bushings for connecting primary to right-angle feeders
- Loadbreak switches installed in transformer cabinet or a closely coupled cabinet
- Tap changing control apparatus [usually a no-load tap changer (NLTC) device that can change the output voltage by about 5%]
- Internal circuit protection devices to open primary line when there are short circuits and severe overloads
- Equipment such as liquid-level gauges, drain valves, radiator guards, sampling devices, and pressure relief valves (for liquid-filled transformers only)
- Internal lightning arresters
- Internal surge arresters for protection against line or switching surges
- Provisions for current and potential transformers and metering
- Future fan provision for such installation at a later date
- Key interlocks or padlocks to coordinate opening of enclosure panels with operation of HV switch
- Provisions for ground-fault detection
- Installation of small control power transformers in cabinet to operate various 120/240 V accessories for medium-voltage transformers
- Seismic bracing for units installed at locations subject to earthquakes

15.4.24 New Techniques of Analysis and Design of DPV-GTs for Photovoltaic Solar Conversion

Flyback inverters for high efficiency and low footprint were recently introduced for distributed transformers [1]. A similar technique but for higer ratings suitable up to 500 KVA energy conversion is described on similar lines of methodology, although this analysis and design technique is yet to be applied for DPV-GTs using IGBT/IGCT silicon devices. As indicated in the references, a prototype is built for a 100 W capacity system. This architecture, as in Figure 15.12, works much more efficiently than the topologies that are implemented with full-bridge and or push-pull configurations. In addition a snubber circuit is used to suppress the transient voltage stress on the switch interconnecting the secondary of the DPV-GT (Figure 15.21).

A parallel connection of flyback converters can be employed to achieve a higher current-carrying capacity as for a 1000 KVA total system capacity if needed. In the primary side further paralleling can be used to share the current so that ECL, copper losses, and core loss due to the incremental

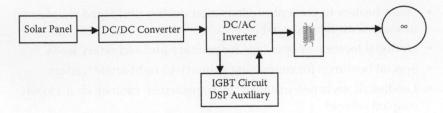

FIGURE 15.21
Architecture of DPV-GTs with improved efficiency.

change in operating flux density (B) can be appropriately calculated and adjusted to satisfy guaranteed losses.

15.4.25 Design of Magnetic Circuit

The crucial part to obtain optimum efficiency is to design the core so that the related core loss is minimized for the DPV-GT. By introducing the air-gap the operating slope of the B/H curve is reduced which results in lower values of residual flux density and an increase in the working range of the new methodology for DPV-GT (Figure 15.22). Air-gaps can be located in the center limb or the outer limbs as shown in Figure 15.14. The former is generally preferred to keep losses in windings close to the air-gaps as low as possible caused due to fringing flux caused by the air-gap. Thus this air-gap is purposely kept low to further reduce the fringing flux. By locating the air-gap in the center limbs, the outside core-limbs are structurally more stable and can sustain larger forces in seismic conditions as required in the accessories described earlier. To reduce the losses caused by proximity to the air-gaps, an interleaving method is used as shown in Figure 15.23.

FIGURE 15.22
(See color insert) Magnetic circuit design with air-gap in the central limb.

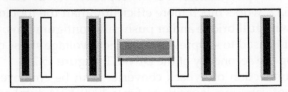

FIGURE 15.23
(See color insert) Windings arranged in interleaved fashion to reduce losses by fringing.

The number of turns in the windings is calculated by using the basic principle of calculating the magneto motive force (mmf) as follows. The flux linkage is given by

$$\chi = N\Phi \tag{15.20}$$

$$N\Phi = LI \tag{15.21}$$

$$B = \Phi/A \tag{15.22}$$

From the above equations the value of the number of turns in each coil is given by

$$L = N^2\rho \tag{15.23}$$

where χ is the flux linkage, N is the number of turns, Φ is the magnetic flux, B is the magnetic flux density, A is the area of cross section of the chosen path, and ρ is the magnetic permeance of the path. In Figure 15.24 the overall closed loop integral of $\int H.dl$ has two magnetic media in its closed loop: one for the magnetic core and the other for the air-gap.

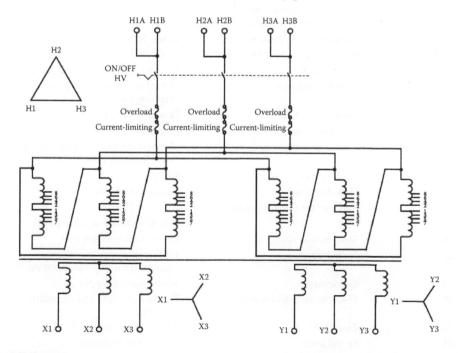

FIGURE 15.24
Typical design specifications and a methodical design procedure.

Datasheet of 1 MVA Transformer Using a Standard Inverter

	Description	Value
1	Output power	Nominal apparent power of 1000 kVA
2	Overload capacity	No
3	Primary voltage (kV)	12.47/13.8/20.6/24.9/27.6/34.5
4	Primary taps	2% ... 2.5% taps above and
		2% ... 2.5% taps below nominal
5	Nominal current, Primary (A)	46.3/41.9/28.1/23.2/21/16.8
6	Secondary voltage (V)	200Y : 200Y
7	Secondary current (A)	1,444
8	Primary BIL (kV)	95/95/125/125/150/150
9	Secondary BIL (kV)	30
10	Vector group	Dy1y1
11	Frequency (Hertz)	60
12	Impedance voltage Z HV-(LV1 + LV2) (%)	5.0 ... 10.0
13	Impedance voltage Z HV-LV1 (%)	
	Base: Half of nominal power (kVA)	4.0 ... 6.6
14	Impedance voltage Z HV-LV2 (%)	
	Base: Half of nominal power (kVA)	4.0 ... 6.6
15	Impedance voltage Z LV1-LV2 (%)	
	Base: Half of nominal power (kVA)	> 9.0
16	No load losses (Watts, @20°C)	1100 ± 10%
17	Short-circuit losses (Watts, @85°C)	9000 ± 10%
18	Tap changer	No load, five-position tap changer
19	Primary bushings	Qty (6) deadbreak, one piece bushings, 600 A
20	Primary configuration	Dead front loop feed
21	Secondary configuration	Live front
22	Load break switching	Three phase, two-position loadbreak switch (200 A at 34.5 kV)
23	Cooling class	KNAN
24	Dielectric fluid	Biodegradable fluid
25	Ambient temperature range	20°C ... +50°C (−40°C available upon request)
26	Temperature rise	65° average winding rise
27	Duty cycle	100% continuous operation; designed for step-up operation
28	Coil material	Aluminum or copper
29	Electrostatic shield winding	Between primary and secondary windings
30	Elevation	1000 m above sea level
31	Sound level	NEMA TR1 Standard

Continued

Datasheet of 1 MVA Transformer Using a Standard Inverter

	Description	Value
32	Secondary bushings	Qty (6) integral aluminum six-hole spade bushings, plus (1) low voltage bushing for the electrostatic shield winding. 1500 A rated each
33	Gauges and fittings	Thermometer, dial type, with alarm contacts
(optional)	Liquid level gauge with alarm contact	
(optional)	Pressure/vacuum gauge with contacts	
Pressure relief device, 50 SCFM		
Drain valve with sampler		
Nitrogen blanket		
34	Overcurrent protection	Overload and current limiting fuses in series
35	Arresters	Not included
36	Certifications	UL listed
37	Coil type	Four-winding

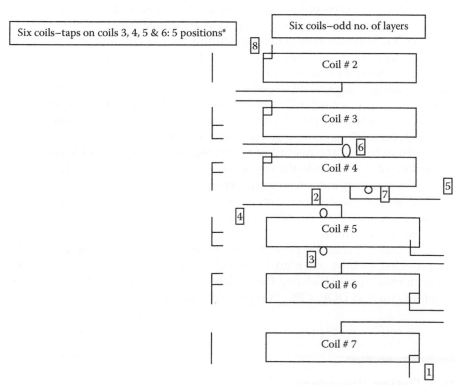

FIGURE 15.25
Typical design specification and a methodical design procedure, taps on coils 3, 4, 5, 6.

The designs of DPV-GTs are generally two windings, three windings with two LVs and one HV, three windings with center-line design on LV; and four windings with two LVs and two HVs in a double-story formation.

Each impedance needs a reference value, from 4% to 8% generally can be as high as 18% to 20% for special dry-type transformers with 2 WDGS dy1, ynd1, ynyo connection.

Accessories and loading 500 kW to 1 MW harmonics above 5% High efficiency Optimized size to fit into skids or mobile platforms Electrostatic shields between lv and HV WDGS For protection from switching activities WDGS designs depends on imp values split if higher imp need Example of solar pv transf Life expectancy 25 yrs Imp4% to 6.6% and about 10% Over voltage.

15.4.26 Impedance, Accessories, and Loading

Each specified transformer impedance needs a reference value from 4% to 8%. Sometimes this impedance can be as high as 18 to 20% for special dry-type transformers as in the case of a winding connection: 2 Windings—dy1, ynd1, ynyo. Aspects of accessories and loading are specified as follows; generally, a loading of 500 kW to 1 MW is expected for transformers with harmonics above 5%. Transformers generally have high efficiency and an optimized size to fit into skids or mobile platforms. Electrostatic shields are provided between LV and HV windings for protection from switching activities. Winding designs depend on impedance values and are split if higher impedance is needed. For a solar PV transformer, the life expectancy is generally 25 years. For a typical transformer, the impedance may be between 4% to 6.6% and about 10% over voltage condition that may cause a serious problem in the grounding of the neutral of the transformer. The no load losses have a tolerance of +10%, while the copper losses have a maximum tolerance of +10%. Winding tappings on an transformer are generally +-10% on the nominal voltage. For the dual winding structure on the HV side, a transformer will need six bushings. All other requirements are as per conventional pad mounted transformers.

A drip tray is generally provided for oil leakage and electrostatic shields are provided to take out high frequency disturbances to satisfy C57.12.80 requirements. Also, efficiency has different levels as specified by different standards such as DOE, CEC, European Weighted CEC, etc.

15.5 Design Procedure

A design procedure for a 1 MVA, three-phase, 50 Hz, OD service, 11-6.6 kV/426 V, Δ/Y, temperature rise of 40/50°C, ONAN/OFAF, with tappings* of +7.5% to –7.5% in steps of 2.5% on HV is illustrated on a design data sheet. The BIL specified is 95 to 60/kV peak; for pf test: –38-28/1 kV rms; with an O/C switch.

15.5.1 Design Process

The entire design starts by first evaluating the volts per turn for a prescribed magnetic flux density B and a chosen core diameter of the transformer for a given core material.

1. The thumb rule value of the volts per turn is calculated such that the volts/turn/square inch = 0.25.

2. The turns at nominal voltage are calculated for each set of windings with maximum flux density of 1.7 T.

3. Care is taken to identify the winding with the tappings. Generally, these are for HV or LV variations, respectively. In this case it is for HV variations as in the specification.

4. Conductors are chosen for each winding with a nominal value of 2.2 A/mm², and a 10% variation is allowable. Any higher value will be tolerable for tapping windings or unloaded windings in the configuration.

5. Paper covering for each winding is chosen such that the voltage withstand requirements are adequately satisfied. In this case this is for a maximum of 11 kV.

6. An appropriate size of the core diameter is chosen to yield the correct cross-sectional area to allow magnetic flux to pass without causing a saturation of the core material and limited to 1.7 T.

7. Electrical specifications are entered as in Table 15.10(a).

8. Calculations for physical dimensions of coils, stray losses, ECL, impedance calculations, temperature rises, and testing schedule are filled out as in Table 15.10(b).

9. The entire design mechanism involves the proper insulation coordination between windings, core, and other electrically active parts of the transformer like the cleat bars, cables, and the tank which is generally considered to be at the ground potential. Care has to be taken to understand the effects of testing the transformer for BIL, power frequency, and induced voltage tests on this ground potential.

10. There are two sets of HV windings at 11 and 6.6 kV, respectively. Tap turns are placed on coils 3, 4, 5, and 6 as shown in Figure 15.25. The winding diagram is as in Figure 15.24.

Method to calculate the other aspects of design are illustrated in the prequel.

TABLE 15.10

Design Data

Part A

IRON	Ckt	Grade M4	No. of bundles	Nett area	B	1.7022T	Weight	VA/lb	kVA
CORE	Dia	Window	Centers		Leg	333.7	Height	Loss	Watts
Radial	Dia.	Ext. Dia.			Leg	339.9	Height		Watts
SHELL	Blank				Tongue	Length	Opening		
Dimensions	217								

Weight		
Legs 333.7	Yokes 339.9	Tot. Wt. 942

SF 0.97

V/T 12.61

Loss: −2.0 × 0.9 (Watts/lb) ×2.0 Legs 333.7 Yokes 339.9

×1.0 ×0.85

Total Loss 1601.4 W

Windings

Windings	Coil #	No. of Coils 1×3	HV	Turns	Current		Sec LV / HV
No. of Legs Wound	H or L	1×3					408 / 13.6/8.3
LV						Den	3.41 / 2.22/3.65
			HV Current	Turns	Current	M.T. Ins.	258.2π / 1.258π
Coil No. 1	6 × 3		2–7	872/523	30.3/50.5	Length m	48.6m / 564/2477
Turns/coil		872(±66)/ 523(±33) HV Taps	1391	2,7/3,4,5,6– six coils			
No. of layers			19	19			
			2 layers	87–4/191–7T		Wt. Kg	92/94 / 68/71–183/189
Turns/layer			10/9	7/9			
Ins. Between layers			Rad = 26.2	13/22			

Conductor					
3/16" spacer		RAD = 37.34/38.74			
2 × 12 + 2 × 11		.164"/.128"			
PC 0.4		.178"/.140			
47.6 × 10.2		PC 14/PC12			
			Res. 75°C	0.002515	0.875/5.78–6.295
			I²R 75°C	4870W	6.06kW@11kV / 5.92kW@66kV
			W/lb 75°C		
			W/sq.in 75°C		

Former					
Tank	875	Transf	1570	Res 75°C	0.0486/ 0.1215
Radiator	325	Tank	1260	I²R 75°C	2.325 kW
OLTC	3500	Radiator	30	W/lb 75°C	
Fill Up/ Pockets	40	Oil OC	860	W/sq.in	

Continued

TABLE 15.10 (Continued)

Design Data

Former	3 mm Cyl
TANK	−1250 × 500
	217/232 +1.5
	1400 - H
Notes:	rectangular, 11 kV Δ
	6.6 kV Δ
Cons/Cabl box	40
Oil Total	1010
Weights	3850
Performance	1.6/14
Fe Cu	1.6/11.7 KW
Total	Guar 1.605/11.7 KW
%Z.	4.97% — Guar 5%
HV Vol	11-6.6KV - Δ 1 — MVA
LV Vol	415 Y — MVA
Phase	3 / 3
Cycles	50 / 50
Service	OD / OD
Cooling	ONAN/OFAF / ONAN/OFAF
Serial No.	06242013
Designer	HMS
Core dia/W/H-Win	217/565/400 mm
Frame	Shields
Customer	37.5/50 ABC Co.
	none — Elec. Spec. 062113

Part B

Radius (mm)	Axial Dimensions and End Distance
Former 217	
Sp/Wrap Rad 7.5-232	60 × 2 = 120
232	80 × 4 = 320
	440
LV 26.2 258.2 π	56
284.4	
LV/HV 10.4–294.8π	= 496
305.2	+ 69
HV 38.74–343.9π	2210.0
382.68	End clearance 69
17.32	Total: 565 (mm)
400	

Stray Loss		HV/LV Reactance @ normal tap		
= 4000 W from calculations as shown in Chapter 14	35	56.4		
	21.4	10.4		20
			56.4	
	18.8	68.8		
	10.4	22.93		
	2	496		
	31.2	518.93		

HV   LV

518.93

$\%X = (1000 \times 31.2 \times 294.8 \times \pi x)/(3 \times 2.54 \times 518.9312.61^2)$

= 4.59% +5% tol = 4.83%

Continued

TABLE 15.10 (*Continued*)

Design Data

	Temperature Rise
Eddy Current Loss	
600 W from calculations as shown in Chapter 14	40/50°C
Clearances on LV Side	**Testing Schedule**

Note:　These are standard national abbreviations in the transformer industry.

References

1. Shertukde, Hemchandra M., Transformer theory and design, class notes for ECE 671 at University of Hartford, Connecticut, 1992 to present, Copyright © Hemchandra Shertukde.
2. Shertukde, Hemchandra M., *Transformer: Theory, design and practice with practical applications*, VDM. Verlag. Dr. Müller, August 8, 2010, Germany.

Chapter 15 Problems

1. A 15/20 MVA 220 kV (±10% variation)/33 kV, Y/Y oil-filled three-phase transformer has the following details:

   LV winding:

   a. ID: 611 mm

   b. Mean diameter: 649 mm

   c. OD: 687 mm

   d. Axial length: 1400 mm

   e. Volts per turn: 82.12

   Radial clearance between LV and tap: 25 mm

   Tap winding:

   a. ID: 737 mm

   b. Mean diameter: 750 mm

   c. OD: 763 mm

   d. Axial length: 1400 mm

   e. Volts per turn: 82.12

   Radial clearance between tap and HV: 80 mm

   HV winding:

   a. ID: 923 mm

   b. Mean diameter: 991 mm

   c. OD: 1059 mm

   d. Axial length: 1400 mm (with 98 discs having 16 turns per disc)

   e. Volts per turn: 82.12

   A. Draw the mmf diagram for the transformer windings at normal tap (i.e., at 100% rated voltage on HV winding).

   B. Calculate the percent reactance of the transformer at the normal tap (i.e., at 100% rated voltage on the HV winding).

2. Complete the following:

    a.   Calculate the outward force on the one conductor of the HV windings of the transformer in Problem 1 in pounds per inch of length.

    b.   If the mismatch in the axial lengths of the HV and LV windings in Problem 1 is about 200 mm due to mechanical misalignment, calculate the axial force on either side of the coils.

3. The transformer in Problem 1 has to be designed with interleaved winding in the HV as compared to ordinary continuous double disc.

    a.   Draw the winding diagrams for the HV disc sections with interleaving as well as double-disc configuration.

    b.   Draw the distribution of the voltage on the HV winding with interleaving and CDD configurations (i.e., voltage versus length of winding for both α).

    c.   What is the improvement in α by the use of interleaving configuration? Give the answer with a numerical value.

16

Special Tests Consideration for a Distributed Photovoltaic Grid Power Transformer

Every DPV-GT transformer that is manufactured undergoes some form of factory testing. For DPV-GTs, these tests are quite extensive and a certain percentage of test failures do occur. Typically around 5% of the transformers produced will fail at least one of these tests. Test requirements are spelled out in a number of industry standards and specifications.

There is some overlap among industry standards with respect to testing. In recent years ANSI/IEEE Standard C57.12.90 has been generally adopted by the other standards for testing liquid-immersed distribution, power, and regulating transformers. Some of the more significant standard factory tests specified in ANSI/IEEE Standard C57.12.90 are itemized in several sections in these standards for DPV-GTs.

These transformers have special requirements of type tests, especially related to partial discharge. These requirements are stringent. Generally, IEEE standards require a PD level of 100 µV at 100% voltage and varying levels for increased voltage levels. DPV transformers demand levels as follows: 125 µV at 110% voltage levels and 135 µV at 125% voltage levels. These voltage levels are experienced at the induced voltage tests conducted on the transformer as described later.

Dielectric test

Switching impulse test

Lightning impulse test

Partial discharge (PD) test with acoustic methods: This test is conducted as per the guidelines of C.57.127—"User's guide for use of acoustic methods for liquid filled transformers." In addition, this PD detection and location methodology has been extended to dry-type DPV-GTs.

Partial discharge levels stipulated by the customer are lower than those specified by IEEE or IEC standards. The new and improved methodology as introduced by Diagnostic Devices Inc. has proved to be extremely successful especially for the first time for dry-type DPV-GTs.

A four-channel PD device using acoustic sensors was used to detect PD for a dry-type 1000 KVA DPV-GT. The testing is done when the transformer is in the plant being tested for induced voltage test. Figure 16.1 shows the

FIGURE 16.1
(See color insert) Four-channel PD diagnostic device. (Courtesy of Diagnostic Devices Inc.)

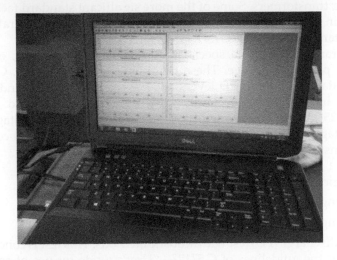

FIGURE 16.2
(See color insert) Data screen of a four-channel PD diagnostic device. (Courtesy of Diagnostic Devices Inc.)

present device used in the market. Figure 16.2 shows the data collected for dry-type DPV-GT with the acoustic sensors mounted for PD detection and eventual location analysis. The PD levels are very low, since these transformers are also used in communication network. Levels are lower than the IEEE standard. In this particular DPV-GT case, the PD required by the customer was as low as 50 pC. Figure 16.3 shows the dry-type transformer under induced voltage test being concurrently tested for partial discharge test using acoustic methods.

FIGURE 16.3
(See color insert) 1600 kVA dry-type DPV-GT being tested with a four-channel briefcase PD diagnostic device. (Courtesy of Diagnostic Devices Inc.)

Other tests include the following:

Insulation power factor

Insulation resistance test

Noise measurement

Heat run or temperature rise test

Short-circuit test

Maximum continuous rating test

Over-current test

DC bus over-voltage test

Anti-islanding test (please refer to Chapter 6 for more details)

Over-/under-voltage/frequency test

Ground-fault test

Voltage and current harmonics test

Power limiting test

Ratio test

No load losses

Load losses and impedance voltage (Tables 16.1 and 16.2)

TABLE 16.1

Typical Guaranteed Values for Standard kVA
Ratings: No Load/Load/%Z

kVA	No Load Loss (W)	Load Loss (W)	%Z
150	550	3050	4.75
250	700	4350	4.75
315	780	5700	4.75
400	890	6600	4.75
500	1080	7800	4.75
630	1500	8900	4.75
750	1390	10100	5.0
1000	1600	13300	5.0

TABLE 16.2

Impedance Voltage for Low Voltage Rating 700 V and Above

Percent Impedance Voltage KVA Rating (Low voltage <700 V)	Impedance		
75	1.10–5.75		
112.5–300	1.40–5.75		
500	1.70–5.75		
750–3750	5.75 nominal		

	Impedance		
Low voltage >700 V	≤150 kV BIL	200 kV BIL	250 kV BIL
1000–5000	5.75	7.00	7.50
7500–10,000	6.50	7.00	7.50

Insulation power factor: This test verifies that vacuum processing
   has thoroughly dried the insulation system to the required limits.
   Generally, in practice, the tan-delta test is used for this evaluation.

Ratio, polarity, and phase relation: This test assures correct winding
   ratios and tap voltages; it also checks insulation of HV, taps, and
   LV circuits. It also checks the entire insulation system to verify all
   live-to-ground clearances.

Resistance: This test verifies the integrity of internal high-voltage and
   low-voltage connections; it also provides data for loss upgrade calculations. Losses are guaranteed and, in practice, failure to meet
   guaranteed losses is subject to penalty as per the guidelines of the
   underlying contract.

Applied potential: Applied to both high-voltage and low-voltage windings, this test stresses the entire insulation system to verify all live-
   to-ground clearances.

TABLE 16.3

Audible Sound Levels for Oil-Filled/Dry-Type DPV-GTs

Average Sound Level (dB)	350 kV BIL and Below: Rating Cooling ON/OW/OFB kVA	350 kV BIL and Below: Rating Dry-Type Ventilated kVA	350 kV BIL and Below: Rating Dry-Type Sealed, kVA
57/62/61	700	700[a]	700[b]
58/64[a]/63[b]	1000	1000[a]	1000[b]
60/65[a]/64[b]	1500	1500[a]	1500[b]

Induced potential: In general this test is conducted at 3.46 times normal plus 1000 V for reduced neutral designs.

Loss test: These design verification tests are conducted to assure that guaranteed loss values are met and that test values are within design tolerances. Tests include no-load loss and excitation current along with impedance voltage and load loss. Guaranteed losses are generally specified within ±5%

Leak test: Pressurizing the tank to 7 psig assures a complete seal, with no weld or gasket leaks, to eliminate the possibility of moisture infiltration or fluid oxidation.

Design performance tests: The design performance tests include the following:

Temperature rise: Our automated heat run facility ensures that any design changes meet ANSI/IEEE temperature rise criteria.

Audible sound level: Ensures compliance with NEMA requirements (Table 16.3).

Lightning impulse or basic impulse level (BIL) test: To assure superior dielectric performance, this test consists of one reduced wave, two chopped waves, and one full wave in sequence, precisely simulating the harshest conditions (Table 16.4).

TABLE 16.4

Ratings for Three-Phase Transformers

(Single Ratio) Primary Voltage	BIL (kV)	Secondary Voltage	BIL (kV)
2400 Delta	60	208Y/120	All 30 kV
4160 Delta	60	480Y/277	
4800 Delta	60	575Y/332	
7200 Delta	75	600Y/347	
12,000 Delta	95	690Y/398	
12,470 Delta	95	240 Delta	
13,200 Delta	95	480 Delta	
13,800 Delta	95	240 Delta with 120 Mid-Tap	
14,400 Delta	95	480 Delta with 240 Mid-Tap	
16,430 Delta	125	See left column for voltages over 700 V	
34,500 Delta	150		
43,800 Delta	250		
4160GrdY/2400	60		
8320GrdY/4800	75		
12,470GrdY/7200	95		
13,200GrdY/7620	95		
13,800GrdY/7970	95		
22,860GrdY/13200	125		
23,900GrdY/13800	125		
24,940GrdY/14400	125		
34,500GrdY/19920	150		
43,800GrdY/25300	250		

Note: For complete connector rating, see ANSI/IEEE 386. Transformers are suitable for connectors with phase-to-ground or phase-to-ground/phase-to-phase high-voltage ratings as listed. Arrester coordination may require higher BIL on multiple connections than indicated to achieve a minimum protection level of 20%.

17

Safety Protection and Shipping and Dispatch for Distributed Photovoltaic Grid Transformers

Distributed photovoltaic grid transformers (DPV-GTs) are used in pad-mount, station, substation, and grounding transformer applications. Thus these transformers require all security measures to be provided for safety in public areas.

Islanding is a condition when a generating source like a PV generator and the corresponding load it provides remains isolated from the main grid. Such DPV grid tie is generally on the low-voltage distribution side, but higher voltages may be involved in the recent ascending voltages where DPV systems find applications. These are generally nonutility generating systems where the utility has no control over its operations. Thus, islanding can affect the operations of such DPV-GTs because some of the islanding measurement systems depend on the voltage, frequency variations, and harmonics that exist due to inherent systems like inverters that enable the conversion of DC to AC in PV systems. A critical and dangerous situation can arise when such an islanding takes place when major or large power transformers are not involved and there are local transformers in the local PV grid system.

Islanding can be extremely dangerous to utility workers, who may not realize that a circuit is still powered, and it may prevent automatic re-connection of devices. The latter thus prevents the control of the utility on the operations of such local distributed generating systems. For that reason, distributed generators must detect islanding and immediately stop producing power; this is referred to as anti-islanding. While still connected to the local DPV-GT, the end user may also be endangered, and this situation can be fatal where normal operations such as connecting or disconnecting from the DPV-GT and the grid tie are not properly coordinated. Thus communications at each end, mainly at the grid and at the PV end, and vice versa, become extremely important. The cybercrimes that have been committed recently also make such requirements about cybersecurity more relevant. Considerable efforts have been made by all manufacturers to strengthen these aspects in their product offerings for the DPV-GT applications.

17.1 Islanding Detection Methods for Safety Monitoring and Control

Some of the most popular islanding detection and measuring systems developed and used to immediately stop generating power or anti-islanding are classified in two categories:

1. Passive methods
   a. Under-/over-voltage
   b. Under-/over-frequency
   c. Voltage phase-jump detection
   d. Harmonics detection
2. Active methods
   a. Impedance measurement
   b. Impedance measurement at a specific frequency
   c. Slip-mode frequency shift
   d. Frequency bias
   e. Utility-based methods
   f. Manual disconnect
   g. Automated disconnect
   h. Transfer trip method
   i. Impedance insertion
   j. SCADA

17.2 Safety, Protection, and Monitoring

An inverter continuously monitors the condition of the grid and in the event of grid failure it automatically switches to off-grid supply within seconds. The solar system is then resynchronized with the grid within a few minutes after restoration of the grid. The SPV system must be suitably earthed in accordance with relevant standards. Over-voltage protection must be provided by using metal-oxide varistors on the DC and AC sides of the inverter.

In addition to disconnection from the grid on no-supply, under-voltage, and over-voltage conditions, PV systems must be provided with adequately rated fuses on the DC and AC sides of the inverter, apart from disconnecting switches. Adequately rated fuses must also be provided in each solar array module to protect them against short circuiting. The inverter's automatic disconnection system may fail and continue to provide electricity to the grid during grid failure. To avoid accidents in such an event, a manual disconnect switch apart from automatic disconnection to the grid must be provided. It must be locked during a planned shutdown. Protections that must also be provided are

- Avoiding battery overcharging
- Avoiding battery overdischarge
- Battery overload protection
- Ground-fault protection systems
- Suitable earthing arrangements

17.2.1 DPV-GT Specific Controls and Related Protections

There are many incidental controls associated with the operations of the DPV-GT especially related to inverters—for example, auto wake-up. Generally, the DPV-GT inverter connected in the single-line diagram should be able to wake-up when the available power is more than the total loss of the available inverter system. If the auto wake-up is premature and before the actual wake-up is intended to be effected, a negative power flow from the grid will take place that will hamper energy production by the PV generator. Similarly a delayed wake-up will cause the energy production with detrimental consequences. As the available power is dependent on the weather conditions, the wake-up algorithm should be adaptive in nature and the wake-up calls adjusted dependent on the insolation number available at the time.

17.2.2 Maximum Power Point Tracking (MPPT)

It should be noted that the current voltage or IV characteristics as in Figure 12.15, with the power axis on the right-hand-side of the figure can affect the power to change as a function of the IV characteristic. The tangent drawn to the IV characteristic represents the power generated in the PV system that is seen by the DPV-GT. This curve changes as the insolation level changes. MPPT has to react fast enough to increase energy extraction but not too fast to cause instability. One can notice that these curves are similar to the torque-slip characteristic of an induction motor. There are stable and unstable regions of operations. Tracking along the left side of the curve should be avoided. At the same time MPPT has to take into account the grid voltage tied to the DPV-GT while tracking lower. This is crucial when the low-voltage ride-through conditions occur.

17.2.3 Protection from DC Bus Over-Voltage

In extremely cold conditions where temperature can fall below zero, the grid and solar ties can fail and cause a DC over-voltage condition. In cases where the DC bus exceeds designated maximum voltage, say for a system of 600 volts, the inverter has to be shut down and the solar array will be disconnected from the DC bus to protect auxiliary equipment like capacitors, disconnects, and wire insulations at specified nominal voltage.

17.2.4 Protection from DC Bus Over-Current

An increased DC bus over-current can be easily caused due to a very low DC bus voltage coupled with higher AC grid voltage which generally is created by enhanced solar insolation conditions. A hardware failure or a short circuit in the DC bus circuitry can cause a subsequent and immediate bus over-current situation. Sometimes a sudden inrush current is sent to the DC bus capacitor which is already connected to a PV array that is generating the voltage. In such conditions the increased inrush current can damage the DPV-GT and the inverter, and the DPV-GT needs to be disconnected and isolated.

17.2.5 Protection from Reverse DC Bus

There are chances that sometimes the inverters are connected in reverse polarity for the DC connection. The controls and protection system should identify this situation and create a warning or go into a sleep mode of operation. If not avoided, this situation can damage the capacitors and cause over-current as indicated above.

17.2.6 Protection from Ground Faults

If a PV cell is grounded through the chassis, this results in a ground fault. The resulting ground-fault current flows through the ground to the negative terminal first and then to the positive terminal via the PV cells. This will be registered by the current sensor in the negative terminal resulting in the shutdown of the inverter, thus disconnecting it from the grid and the DPV-GT as well. This ground-fault current should be detected before it affects the working of the DPV-GT.

The above will affect grid safety and the safety of the DPV-GT, resulting in possible islanding, a single-phase open condition, AC over-voltage, AC under-voltage, an over-frequency condition, an under-frequency condition, an AC over-current condition, and short-circuit conditions that in turn will affect the stable operation of the DPV-GT. Conventional methods to isolate and tackle such conditions are already available with the traditional transformers. Similar protective schemes can be applied to DPV-GTs.

17.3 Potential Operations and Management (O&M) Issues

Sixteen potential failures or damages that can occur on a solar farm and how they could impact operations if not addressed in a timely manner are as follows:

1. Perimeter fence damage: Damage caused to the perimeter fence can immediately have a negative effect on facility operations. Whether the damage was due to vandals, a storm, or even an animal, this is an item that needs immediate attention. Not only can people be injured due to the high voltage produced by the system, but the expensive equipment is at risk if intruders enter the area with intent to destroy or steal items. Regular inspection and quick response to this is crucial for all solar farms.

2. Ground erosion: A naturally occurring process in nature, soil and ground erosion are caused by water and wind. This is expected as a gradual occurrence and planned for at a certain periodic rate, but sudden erosion can have a deleterious effect on a PV plant. Loss of topsoil can lead to reshaping of the ground and the creation of channels, holes, and slopes in earth. This could cause racking to shift, affecting the ability of panels to generate the energy. It could also lead to flooding and destruction of equipment. Proper and frequent site monitoring will alert asset managers to anything out of the ordinary happening that could put operations at risk.

3. Transformer leakage: Routine maintenance that certifies that transformers are in good condition every year helps avoid transformer leakage. A transformer leak can cause land contamination and other safety risks. Knowing if a leak is present and planning for maintenance to repair or replace it can be key in keeping energy generation at a maximum. There several ways to carry out preventive maintenance in transformers; however, monitoring transformer oil temperature, pressure, and level to prevent a transformer from leaking in the first place is the best way to avoid downtime issues. To prevent fatal errors, a parameter range is set and automatic alarms can be issued to check on site before the problem scales.

4. Various inverter damage: Taking the low-voltage, high-current signals from PV panels and converting into the voltage compatible with the utility grid, inverters are core components of grid-connected systems. Monitoring of inverters is of high importance, because changes to voltage and frequency may occur that affect performance as well as the safety of those in proximity. Inverter damage may lead to the complete failure of the PV plant or partial string outages as a result of defective inverters. Inverter failures are responsible for

roughly 80% of PV system downtime. Clearly a response to any inverter damage must be taken quickly.

5. Broken conduit: A broken conduit poses the danger of shock as well as chaos on the operating system as charges are uncontained. When the construction of a site is finished and the plant goes into operation, earth movements may happen as the ground stabilizes. These movements can cause broken conduit and other issues with cables. Measuring isolation on cables ensures underground runs are damage free. This is important because broken conduit can cause a cable to break or damage the insulation which can cause a fire and personal hazards.

6. Combiner box damage: With the ability to simplify wiring, combiner boxes combine inputs from multiple strings of solar panels into one output circuit. Normally 4 to 12 strings are connected to a combiner box. If damaged, they pose a safety risk as well as a major decrease in productivity.

7. Vegetation overgrowth: Vegetation can transform from a benign nuisance into a major issue very quickly. In addition to attracting animals that then cause their own brand of destruction, vegetation can shade cells, interfere with wiring, and affect structural integrity.

8. Cell browning/discoloring: In addition to providing power, UV radiation will lead to aging in panel cells, seen as browning and discoloration. This degradation in the film leads to impaired output and productivity.

9. Panel shading: When designing a PV plant, it is critical that trees and other obstructions are cleared. PV cell electrical output is very sensitive to shade. If shaded, cells do not add to the power produced by the panel, but they absorb it. A shaded cell has a much greater reverse voltage compared to the forward voltage of an illuminated one—it can absorb the power of many cells in the string and the output will fall drastically. Removal of any trees or structures causing shading will help optimize power output.

10. Shorted cell: A shorted cell can impact productivity if not addressed in a timely manner. Production defects in semiconducting material often go undetected before PV cells are put into solar panel assemblies. Identifying these defects through testing via infrared imaging has been used for more than a decade. This efficient, cost-effective test and measurement methods for characterizing a cell's performance and its electronic structure help ensure maximum energy production.

11. Natural damage: A hailstorm or hurricane can wreak havoc on a solar power plant. Damaged panels or wind-torn racking and other equipment can severely decrease output or completely put a system out of commission. Keeping a pulse on the severe weather and

inspecting the equipment following a storm are necessary for the overall health of the solar farm.

12. Vandalism damage: Vandals pose a major threat to any PV facility. Whether they are stealing or destroying wiring, panels, or other equipment, system damage can occur. A solar farm in North Carolina had golf ball damage by a neighbor who decided to use the array as the 18th hole. Detecting this damage through the use of solar monitoring equipment minimized outages and losses.

13. Defective tracker: An exceptional tool to enhance early morning and late afternoon performance, trackers can increase total power produced by about 20% to 25% for a single-axis tracker and about 30% or more for a dual-axis tracker. Defective trackers can contribute significantly to lowered performance output and should be serviced as soon as detected.

14. Racking erosion: Eroding structures can be a nightmare for a PV facility. Once the structural integrity is degraded, risks to proper water and wind flow within the facility are elevated which can gravely impact the functioning of the facility. As racking moves, panels are moved from their optimal positioning and energy generation suffers.

15. Unclean panels: Dust, snow, pollen, leaf fragments, and even bird droppings can absorb sunlight on the surface of a panel, reducing the light that reaches the cells. Clean surfaces result in increased output performance over the life span of the equipment. Routine cleaning should be a part of all O&M plans.

16. Animal nuisance: No matter whether an animal burrows under a perimeter fence, jumps over it, or goes right through it, animals need to be kept out of a solar farm. Once inside the perimeter, they seem to have a way of finding wires to chew and unknowingly destroy equipment.

17.4 Solar Power Wiring Design

This section covers solar power wiring design and is intended to familiarize engineers and system integrators with some of the most important aspects related to personnel safety and hazards associated with solar power projects.

Residential and commercial solar power systems, up until a decade ago, because of a lack of technology maturity and higher production costs, were excessively expensive and did not have sufficient power output efficiency to justify a meaningful return on investment. Significant advances in solar cell research and manufacturing technology have recently rendered solar power installation a viable means of electric power cogeneration in residential and commercial projects. As a result of solar power rebate programs available throughout the United States, Europe, and most industrialized countries, solar power industries have flourished and expanded their production capacities in the past ten years and are currently offering reasonably cost-effective products with augmented efficiencies.

In view of constant and inevitable fossil fuel–based energy cost escalation and availability of worldwide sustainable energy rebate programs, solar power because of its inherent reliability and longevity, has become an important contender as one of the most viable power cogeneration investments afforded in commercial and industrial installations.

In view of the newness of the technology and constant emergence of new products, installation and application guidelines controlled by national building and safety organizations such as the National Fire Protection Association, which establishes the guidelines for the National Electrical Code (NEC), have not been able to follow up with a number of significant matters related to hazards and safety prevention issues.

In general, small-size solar power system wiring projects, such as residential installations commonly undertaken by licensed electricians and contractors who are trained in safety installation procedures, do not represent a major concern. However, large installations where solar power produced by photovoltaic arrays generates several hundred volts of DC power require exceptional design and installation measures.

17.5 Solar Power System Wiring

An improperly designed solar power system in addition to being a fire hazard can cause very serious burns and in some instances result in fatal injury. Additionally, an improperly designed solar power system can result in a significant degradation of power production efficiency and minimize the return on investment.

Some significant issues related to inadequate design and installation include improperly sized and selected conductors, unsafe wiring methods, inadequate over-current protection, unrated or underrated choice of circuit breakers, disconnect switches, system grounding, and numerous other issues that relate to safety and maintenance.

At present the NEC in general covers various aspects of photovoltaic power generation systems; however, it does not cover special application and safety issues. For example, in a solar power system a deep-cycle battery backup with a nominal 24 V and 500 Ah can discharge thousands of amperes of current if short circuited. The enormous energy generated in such a situation can readily cause serious burns and fatal injuries. Unfortunately most installers, contractors, electricians, and even inspectors who are familiar with the NEC most often do not have sufficient experience and expertise with DC power system installation, and as such, requirements of the NEC are seldom met.

Another significant point that creates safety issues is related to material and components used, which are seldom rated for DC applications. Electrical engineers and solar power designers who undertake solar power system installations of 10 kWh or more (nonpackaged systems) are recommended to review 2005 NEC Section 690 and the suggested solar power design and installation practices report issued by Sandia National Laboratories.

To prevent the design and installation issues discussed, system engineers must ensure that all material and equipment used are approved by Underwriters Laboratories. All components such as over-current devices, fuses, and disconnect switches are DC rated. Upon completion of installation, the design engineer should verify, independently of the inspector, whether the appropriate safety tags are permanently installed and attached to all disconnect devices, collector boxes, and junction boxes, and verify if system wiring and conduit installation comply with NEC requirements. The recognized materials and equipment testing organizations that are generally accredited in the United States and Canada are Underwriters Laboratories (UL), Canadian Standards Association (CSA), and Testing Laboratories (ETL), all of which are registered trademarks that commonly provide equipment certification throughout the North American continent. Note that the NEC, with the exception of marine and railroad installation, covers all solar power installations, including stand-alone, grid-connected, and utility-interactive cogeneration systems. As a rule, the NEC covers all electrical system wiring and installations and in some instances has overlapping and conflicting directives that may not be suitable for solar power systems, in which case Article 690 of the code always takes precedence.

In general, solar power wiring is perhaps considered one of the most important aspects of the overall systems engineering effort, and as such, it should be understood and applied with due diligence. As mentioned earlier, undersized wiring or a poor choice of material application cannot only diminish system performance efficiency but can also create a serious safety hazard for maintenance personnel.

Take for example this wiring design example. Assume that the short-circuit current I_{sc} from a PV array is determined to be 40 A. The calculation should be as follows:

1. PV array current derating = $40 \times 1.25 \times 50$ A.
2. Over-current device fuse rating at 75°C = $50 \times 1.25 \times 62.5$ A.
3. Cable derating at 75°C = $50 \times 1.25 \times 62.5$.

Using NEC Table 310-16, under the 75°C column, we find a cable AWG #6 conductor that is rated for 65-A capacity. Because of ultraviolet (UV) exposure, XHHW-2 or USE-2 type cable, which has a 75-A capacity, should be chosen. Incidentally, the "2" is used to designate UV exposure protection. If the conduit carrying the cable is populated or filled with four to six conductors, it is suggested, as previously, by referring to NEC Table 310-15(B)(2)(a), that the conductors be further derated by 80%. At an ambient temperature of 40 to 45°C, a derating multiplier of 0.87 is also to be applied: 75 A $\times$ 0.87 = 52.2 A. Since the AWG #6 conductor chosen with an ampacity of 60 is capable of meeting the demand, it is found to be an appropriate choice.

4. By the same criteria the closest over-current device, as shown in NEC Table 240.6, is 60 A; however, since in step 2 the over-current device required is 62.5 A, the AWG #6 cable cannot meet the rating requirement. As such, an AWG #4 conductor must be used. The chosen AWG #4 conductor under the 75°C column of Table 310-16 shows an ampacity of 95.

If we choose an AWG #4 conductor and apply conduit fill and temperature derating, then the resulting ampacity is $95 \times 0.8 \times 0.87 = 66$ A; therefore, the required fuse per NEC Table 240-6 will be 70 A. Conductors that are suitable for solar exposure are listed as THW-2, USE-2, and THWN-2 or XHHW-2. All outdoor installed conduits and wireways are considered to be operating in wet, damp, and UV-exposed conditions. As such, conduits should be capable of withstanding these environmental conditions and are required to be of a thick-wall type such as rigid galvanized steel (RGS), intermediate metal conduit (IMC), thin-wall electrical metallic (EMT), or schedule 40 or 80 polyvinyl chloride (PVC) nonmetallic conduits.

For interior wiring, where the cables are not subjected to physical abuse, special NEC code approved wires must be used. Care must be taken to avoid installation of underrated cables within interior locations such as attics where the ambient temperature can exceed the cable rating. Conductors carrying DC current are required to use color coding recommendations as stipulated in Article 690 of the NEC. Red wire or any color other than green and white is used for positive conductors, white for negative, green for equipment grounding, and bare copper wire for grounding. The NEC allows nonwhite grounded wires.

17.6 Solar Power System Design Considerations

Generally, wiring such as USE-2 and UF-2 that are sized #6 or above are to be identified with a white tape or marker. As mentioned in Chapter 12, all PV array frames, collector panels, disconnect switches, inverters, and metallic enclosures should be connected together and grounded at a single service grounding point.

17.7 Shipping and Dispatch Considerations for a DPV Grid Power Transformer

DPV-GTs are generally dispatched palletized if they are less than 500 KVA and on flatbed trucks or well-shaped trucks if greater than 500 kVA and up to 10 MVA. The latter are generally for CSP types of alternate energy systems.

DPV-GTs are mostly dispatched on suitably sized pallets. Sometimes manufacturers like Power Distribution Inc., dba Onyx Power in California, use completely enclosed wooden boxes depending on where they are being shipped and/or customer preferences. These DPV-GTs suitable for solar (inverter-type) transformers are also sealed with plastic wraps to avoid damage from dust/moisture and are safely secured onto pallets/box-bases using steel straps. The transformer is mounted on wooden planks for temporary shipment and the fixture is then placed on pallets, bolted, and enclosed further in a wooden crate [2]. A picture of a recently manufactured dry-type DPV-GT suitable for solar (inverter-type) application is shown in Figure 17.1.

FIGURE 17.1
DPV-GT for solar (inverter-type) application with complicated secondary winding lead connections for the delta/wye configuration in a frontal view. (Courtesy Power Distribution, Inc., dba Onyx Power.)

FIGURE 17.2
DPV-GT for solar (inverter-type) application with complicated secondary winding lead connections for the delta/wye configuration in a side view. (Courtesy Power Distribution, Inc., dba Onyx Power.)

Figures 17.1 and 17.2 show a smaller-size transformer with basic wooden mounts at the base on which the transformer can be easily mounted and then delivered on a flatbed truck with additional straps for anchoring to make it safe for transportation.

Figures 17.3 and 17.4 show front and side views of a solar transformer ready for dispatch after those have been certified and have passed all tests as specified by the end user. The transformer is laid on the palletized foundation as seen from the wooden base in Figures 17.3 and 17.4 to be then covered by a wooden bell-type cover for final delivery on a flatbed truck.

References

1. NEC Table 240-6.
2. Standard operating instructions, Power Distribution, Inc., dba Onyx Power, 2013.

FIGURE 17.3
DPV-GT for solar application. Note the simple winding configuration on the secondary of the delta/wye neutral grounded configuration in a frontal view. (Courtesy Power Distribution, Inc., dba Onyx Power.)

FIGURE 17.4
DPV-GT for solar application. Note the simple winding configuration on the secondary of the delta/wye neutral grounded configuration in a side view. (Courtesy Power Distribution, Inc., dba Onyx Power.)

Appendix A

MATLAB® Program for a Three-Limb Core Design [16]

```
%This program calculates the core sections for a 3MVA,3phase,
   436 volts/11KV Y-Y transformer
%Reads yoke data values from an exel file that was obtained by using
%microstation (program similar to autocad)
a = xlsread('99-yoke (bottom half)with ducts.xlsx');%reads data from
   an xls file
xyoke = a(:,1);%assign first column to x
yyoke = a(:,2);%assign second column to y
for i = 0:26
    d = 2+(4*i);
    b = 1+(4*i);
    c = 3+(4*i);
    l = xyoke(d)-xyoke(b);
    h = yyoke(c)-yyoke(d);
    Lengthyoke(i+1) = l;%stores yoke length for the cross sections
    Widthyoke(i+1) = h;%stores yoke widths for the cross sections
end

Lengthyoke;
Widthyoke;


%Reads leg data values from an exel file that was obtained by using
%microstation (program similar to autocad)
aa = xlsread('99-limb (bottom half)with ducts.xlsx');
%reads data from an xls file
xleg = aa(:,1);%assign first column to x
yleg = aa(:,2);%assign second column to y
for j = 0:26
    e = 2+(4*j);
    f = 1+(4*j);
    g = 3+(4*j);
    ll = xleg(e)-xleg(f);
    hh = yleg(g)-yleg(e);
    Lengthleg(j+1) = ll;%stores leg length for the cross sections
    Widthleg(j+1) = hh;%stores leg widths for the cross sections
end

Lengthleg;
Widthleg;
```

```
%main leg calculations
legcentertopleftaandb = 1700;%section a and b
legcentertopleftcdef = sqrt(Lengthlegyoke(j+1)^2+(0.5*Lengthleg(j+1)^2));
   %section c, d,e, and f
legcentertopleftangleb = 90+atan(Lengthlegyoke(j+1)
   /(0.5*Lengthleg(j+1)))*180/pi;%angle b
legcentertopleftanglea = atan(0.5*Lengthleg(j+1)
   /(Lengthlegyoke(j+1)))*180/pi;%angle a
legcenteraandb(j+1) = legcentertopleftaandb;%Calculates length a and b
   of center leg
legcentercdef(j+1) = legcentertopleftcdef;%Calculates length a,b,c,
   and d of center leg
legcenterangleb(j+1) = legcentertopleftangleb;%Calculates cutting
   angle b of center leg
legcenteranglea(j+1) = legcentertopleftanglea;%Calculates cutting
   angle a of center leg

%main yoke calculations
yokecentertoplefta = 1500+(0.5*Lengthleg(j+1))+abs((Lengthyoke(j+1)
   -Lengthlegyoke(j+1))/(tan(atan(Lengthlegyoke(j+1)/(0.5*Lengthleg(j+1))))));
   %section a
yokecentertopleftb = 1500;%section b
yokecentertopleftc = abs(Lengthlegyoke(j+1)
   /(sin(atan(Lengthlegyoke(j+1)/(0.5*Lengthleg(j+1))))))
   +abs((Lengthyoke (j+1)-Lengthlegyoke(j+1))
   /(sin(atan(Lengthlegyoke(j+1)/(0.5*Lengthleg (j+1))))));%section c
yokecentertopleftd = sqrt(Lengthlegyoke(j+1)^2+(0.5*Lengthleg(j+1)^2));
   %section d
yokecentertoplefte = sqrt((Lengthyoke(j+1)-Lengthlegyoke(j+1))^2
   +(0.5*Lengthleg(j+1))^2);%section e
yokecentertopleftanglea = abs(asin((Lengthyoke(j+1)
   -Lengthlegyoke(j+1))/(sqrt(Lengthyoke(j+1)-Lengthlegyoke(j+1))^2
   +(0.5*Lengthleg(j+1))^2))*180/pi)+abs(atan(Lengthlegyoke(j+1)
   /(0.5*Lengthleg(j+1)))*180/pi;%angle a
yokecentertopleftangleb = 90+abs(atan((0.5*Lengthleg(j+1))
   /(Lengthyoke(j+1)-Lengthlegyoke(j+1))))*180/pi;%angle b
yokecentertopleftanglec = 90-abs(atan(Lengthlegyoke(j+1)
   /(0.5*Lengthleg(j+1)))*180/pi);%angle c
yokecentertopleftangled = 90+abs(atan((0.5*Lengthleg(j+1))
   /Lengthlegyoke(j+1)))*180/pi;%angle d
yokecentertopleftanglee = 90+abs(atan((0.5*Lengthleg(j+1))
   /Lengthlegyoke(j+1))*180/pi);%angle e
yokecentera(j+1) = yokecentertoplefta;
yokecenterb(j+1) = yokecentertopleftb;
yokecenterc(j+1) = yokecentertopleftc;
yokecenterd(j+1) = yokecentertopleftd;
yokecentere(j+1) = yokecentertoplefte;
yokecenteranglea(j+1) = yokecentertopleftanglea;
yokecenterangleb(j+1) = yokecentertopleftangleb;
yokecenteranglec(j+1) = yokecentertopleftanglec;
yokecenterangled(j+1) = yokecentertopleftangled;
yokecenteranglee(j+1) = yokecentertopleftanglee;

end
```

```
%Main Legs
legcenteraandb;
legcentercdef;
legcenterangleb;
legcenteranglea;

%Main Yokes
yokecentera;
yokecenterb;
yokecenterc;
yokecenterd;
yokecentere;
yokecenteranglea;
yokecenterangleb;
yokecenteranglec;
yokecenterangled;
yokecenteranglee;

%Plotting the Yoke cross section
figure('Name','Main Yoke Cross-section')
grid on
shx = 0.0;
shy = 0.0;
sum1 = 0.0;
sum2 = 0.0;
for i = 27:-1:1
u = zeros(27,1);
axis([0 550 -300 300]);
sum2 = sum2+Widthyoke(i);
line([u(i)+shx Lengthyoke(i)+shx],[u(i)+shy u(i)+shy],'Marker','.',
    'LineStyle','-')
line([u(i)+shx Lengthyoke(i)+shx],[Widthyoke(i)+sum1 Widthyoke(i)+sum1],
    'Marker','.','LineStyle','-')
line([Lengthyoke(i)+shx Lengthyoke(i)+shx],[sum1 sum2],'Marker','.',
    'LineStyle','-')
line([u(i)+shx u(i)+shx],[sum1 sum2],'Marker','.','LineStyle','-')
line([u(i)+shx Lengthyoke(i)+shx],[-u(i)-shy -u(i)-shy],'Marker','.',
    'LineStyle','-')
line([u(i)+shx Lengthyoke(i)+shx],[-Widthyoke(i)-sum1 -Widthyoke(i)-sum1],
    'Marker','.','LineStyle','-')
line([Lengthyoke(i)+shx Lengthyoke(i)+shx],[-sum1 -sum2],'Marker','.',
    'LineStyle','-')
line([u(i)+shx u(i)+shx],[-sum1 -sum2],'Marker','.','LineStyle','-')

if i = =1
return
else
shx = shx+(Lengthyoke(i)-Lengthyoke(i-1))/2;
end
shy = shy+Widthyoke(i);
sum1 = sum1+Widthyoke(i);
end
```

```
%Plotting the Leg cross section
figure('Name','Main Leg Cross-section')
grid on
shx = 0.0;
shy = 0.0;
sum1 = 0.0;
sum2 = 0.0;
for i = 27:-1:1
u = zeros(27,1);
axis([0 950 -500 500]);
sum2 = sum2+Widthleg(i);
line([u(i)+shx Lengthleg(i)+shx],[u(i)+shy u(i)+shy],'Marker','.',
   'LineStyle','-')
line([u(i)+shx Lengthleg(i)+shx],[Widthleg(i)+sum1 Widthleg(i)+sum1],
   'Marker','.','LineStyle','-')
line([Lengthleg(i)+shx Lengthleg(i)+shx],[sum1 sum2],'Marker','.',
   'LineStyle','-')
line([u(i)+shx u(i)+shx],[sum1 sum2],'Marker','.','LineStyle','-')
line([u(i)+shx Lengthleg(i)+shx],[-u(i)-shy -u(i)-shy],'Marker','.',
   'LineStyle','-')
line([u(i)+shx Lengthleg(i)+shx],[-Widthleg(i)-sum1 -Widthleg(i)-sum1],
   'Marker','.','LineStyle','-')
line([Lengthleg(i)+shx Lengthleg(i)+shx],[-sum1 -sum2],'Marker','.',
   'LineStyle','-')
line([u(i)+shx u(i)+shx],[-sum1 -sum2],'Marker','.','LineStyle','-')

if i = =1
return
else
shx = shx+(Lengthleg(i)-Lengthleg(i-1))/2;
end
shy = shy+Widthleg(i);
sum1 = sum1+Widthleg(i);
end

%Calculating number of laminatios per each section. Lamination
%thickness = 0.23mm for end legs and end yokes, 0.27mm for yokes,
%and 0.35 mm for limbs

numoflamiyoke = round(Widthyoke/0.285);
numoflamileg = round(Widthleg/0.5);
numoflamilegyoke = round(Widthlegyoke/0.215);

%Writing results
xlswrite('results970.xlsx',Lengthyoke','Main Yoke','A3')
xlswrite('results970.xlsx',Widthyoke','Main Yoke','B3')
xlswrite('results970.xlsx',numoflamiyoke','Main Yoke','C3')
xlswrite('results970.xlsx',yokecentera','Main Yoke','D3')
xlswrite('results970.xlsx',yokecenterb','Main Yoke','E3')
xlswrite('results970.xlsx',yokecenterc','Main Yoke','F3')
xlswrite('results970.xlsx',yokecenterd','Main Yoke','G3')
xlswrite('results970.xlsx',yokecentere','Main Yoke','H3')
xlswrite('results970.xlsx',yokecenteranglea','Main Yoke','I3')
xlswrite('results970.xlsx',yokecenterangleb','Main Yoke','J3')
xlswrite('results970.xlsx',yokecenteranglec','Main Yoke','K3')
xlswrite('results970.xlsx',yokecenterangled','Main Yoke','L3')
xlswrite('results970.xlsx',yokecenteranglee','Main Yoke','M3')
```

```
xlswrite('results970.xlsx',Lengthleg,'Main Leg','A3')
xlswrite('results970.xlsx',Widthleg,'Main Leg','B3')
xlswrite('results970.xlsx',numoflamileg,'Main Leg','C3')
xlswrite('results970.xlsx',legcenteraandb,'Main Leg','D3')
xlswrite('results970.xlsx',legcenteraandb,'Main Leg','E3')
xlswrite('results970.xlsx',legcentercdef,'Main Leg','F3')
xlswrite('results970.xlsx',legcentercdef,'Main Leg','G3')
xlswrite('results970.xlsx',legcentercdef,'Main Leg','H3')
xlswrite('results970.xlsx',legcentercdef,'Main Leg','I3')
xlswrite('results970.xlsx',legcenteranglea,'Main Leg','J3')
xlswrite('results970.xlsx',legcenterangleb,'Main Leg','K3')


xlswrite('results970.xlsx',Lengthlegyoke,'End Yoke','A3')
xlswrite('results970.xlsx',Widthlegyoke,'End Yoke','B3')
xlswrite('results970.xlsx',numoflamilegyoke,'End Yoke','C3')
xlswrite('results970.xlsx',yokesa,'End Yoke','D3')
xlswrite('results970.xlsx',yokesb,'End Yoke','E3')
xlswrite('results970.xlsx',yokesc,'End Yoke','F3')
xlswrite('results970.xlsx',yokesd,'End Yoke','G3')
xlswrite('results970.xlsx',yokesanglea,'End Yoke','H3')
xlswrite('results970.xlsx',yokesangleb,'End Yoke','I3')
xlswrite('results970.xlsx',yokesanglec,'End Yoke','J3')
xlswrite('results970.xlsx',yokesangled,'End Yoke','K3')


xlswrite('results970.xlsx',Lengthlegyoke,'End Leg','A3')
xlswrite('results970.xlsx',Widthlegyoke,'End Leg','B3')
xlswrite('results970.xlsx',numoflamilegyoke,'End Leg','C3')
xlswrite('results970.xlsx',legsa,'End Leg','D3')
xlswrite('results970.xlsx',legsb,'End Leg','E3')
xlswrite('results970.xlsx',legsc,'End Leg','F3')
xlswrite('results970.xlsx',legsd,'End Leg','G3')
xlswrite('results970.xlsx',legsanglea,'End Leg','H3')
xlswrite('results970.xlsx',legsangleb,'End Leg','I3')
xlswrite('results970.xlsx',legsanglec,'End Leg','J3')
xlswrite('results970.xlsx',legsangled,'End Leg','K3')


%Plot of core sections
for i = 27:-1:1
u = zeros(27,1);
axis([-100 9000 -200 3200]);

%left end bottom yoke
line([u(i) yokesa(i)],[u(i) u(i)],'Marker','.','LineStyle','-')
line([u(i)+Lengthlegyoke(i) yokesb(i)+Lengthlegyoke(i)],[Lengthlegyoke(i)
   Lengthlegyoke(i)],'Marker','.','LineStyle','-')
line([u(i) Lengthlegyoke(i)],[u(i) Lengthlegyoke(i)],'Marker','.',
   'LineStyle','-')
line([yokesa(i) yokesb(i)+Lengthlegyoke(i)],[u(i) Lengthlegyoke(i)],
   'Marker','.','LineStyle','-')

%left end leg
line([u(i) u(i)],[u(i) legsa(i)],'Marker','.','LineStyle','-')
line([Lengthlegyoke(i) Lengthlegyoke(i)],[u(i)+Lengthlegyoke(i) legsb(i)
   +Lengthlegyoke(i)],'Marker','.','LineStyle','-')
```

```
line([u(i) Lengthlegyoke(i)],[u(i) Lengthlegyoke(i)],'Marker','.',
  'LineStyle','-')
line([u(i) Lengthlegyoke(i)],[legsa(i) legsb(i)+Lengthlegyoke(i)],
  'Marker','.','LineStyle','-')

%left end upper yoke
line([u(i) yokesa(i)],[u(i)+legsa(i) u(i)+legsa(i)],'Marker','.',
  'LineStyle','-')
line([u(i)+Lengthlegyoke(i) yokesb(i)+Lengthlegyoke(i)],
  [legsa(i)-Lengthlegyoke(i) legsa(i)-Lengthlegyoke(i)],'Marker','.',
  'LineStyle','-')
line([u(i) Lengthlegyoke(i)],[legsa(i) legsb(i)+Lengthlegyoke(i)],
  'Marker','.','LineStyle','-')
line([yokesa(i) yokesb(i)+Lengthlegyoke(i)],[u(i)+legsa(i) Lengthlegyoke(i)
  +legsb(i)],'Marker','.','LineStyle','-')

%Left center leg
line([yokesa(i) Lengthlegyoke(i)+yokesb(i)],[u(i) Lengthlegyoke(i)],
  'Marker','.','LineStyle','-')
line([Lengthlegyoke(i)+yokesb(i) Lengthlegyoke(i)+yokesb(i)],
  [Lengthlegyoke(i) Lengthlegyoke(i)+legsb(i)],'Marker','.',
  'LineStyle','-')
line([Lengthlegyoke(i)+yokesb(i) yokesa(i)],[Lengthlegyoke(i)+legsb(i)
  legsa(i)],'Marker','.','LineStyle','-')
line([yokesa(i) yokesa(i)+0.5*Lengthleg(i)],[legsa(i) Lengthlegyoke(i)
  +legsb(i)],'Marker','.','LineStyle','-')
line([yokesa(i)+0.5*Lengthleg(i) yokesa(i)+0.5*Lengthleg(i)],
  [Lengthlegyoke(i)+legsb(i) Lengthlegyoke(i)],'Marker','.',
  'LineStyle','-')
line([yokesa(i)+0.5*Lengthleg(i) yokesa(i)],[Lengthlegyoke(i) u(i)],
  'Marker','.','LineStyle','-')

%left bottom center yoke
line([yokesa(i)+0.5*Lengthleg(i) yokesa(i)],[Lengthlegyoke(i) u(i)],
  'Marker','.','LineStyle','-')
line([yokesa(i) yokesa(i)+0.5*Lengthleg(i)],[u(i) Lengthlegyoke(i)
  -Lengthyoke(i)],'Marker','.','LineStyle','-')
line([yokesa(i)+0.5*Lengthleg(i) yokesa(i)+0.5*Lengthleg(i)
  +yokecentera(i)],[Lengthlegyoke(i)-Lengthyoke(i)
  Lengthlegyoke(i)-Lengthyoke(i)],'Marker','.','LineStyle','-')
line([yokesa(i)+0.5*Lengthleg(i)+yokecentera(i) yokesa(i)
  +0.5*Lengthleg(i)+yokecenterb(i)],[Lengthlegyoke(i)-Lengthyoke(i)
  Lengthlegyoke(i)],'Marker','.','LineStyle','-')
line([yokesa(i)+0.5*Lengthleg(i)+yokecenterb(i) yokesa(i)
  +0.5*Lengthleg(i)],[Lengthlegyoke(i) Lengthlegyoke(i)],'Marker','.',
  'LineStyle','-')

%left upper center yoke
line([yokesa(i) yokesa(i)+0.5*Lengthleg(i)],[legsa(i) Lengthlegyoke(i)
  +legsb(i)],'Marker','.','LineStyle','-')
line([yokesa(i)+0.5*Lengthleg(i) yokesa(i)+0.5*Lengthleg(i)
  +yokecenterb(i)],[Lengthlegyoke(i)+legsb(i) Lengthlegyoke(i)
  +legsb(i)], 'Marker','.','LineStyle','-')
line([yokesa(i)+0.5*Lengthleg(i)+yokecenterb(i) yokesa(i)
  +0.5*Lengthleg(i)+yokecentera(i)],[Lengthlegyoke(i)+legsb(i)
```

```
      legsa(i)+Lengthyoke(i)-Lengthlegyoke(i)],'Marker','.','LineStyle',
      '-')
line([yokesa(i)+0.5*Lengthleg(i)+yokecentera(i) yokesa(i)
   +0.5*Lengthleg(i)],[legsa(i)+Lengthyoke(i)-Lengthlegyoke(i)
   Lengthlegyoke(i)+legsb(i)+Lengthyoke(i)],'Marker','.','LineStyle','-')
line([yokesa(i)+0.5*Lengthleg(i) yokesa(i)],[Lengthlegyoke(i)+legsb(i)
   +Lengthyoke(i) legsa(i)],'Marker','.','LineStyle','-')


%center leg
line([yokesa(i)+0.5*Lengthleg(i)+yokecenterb(i)+0.5*Lengthleg(i) yokesa(i)
   +0.5*Lengthleg(i)+yokecenterb(i)],[2*Lengthlegyoke(i)+legsb(i)
   Lengthlegyoke(i)+legsb(i)],'Marker','.','LineStyle','-')
line([yokesa(i)+0.5*Lengthleg(i)+yokecenterb(i) yokesa(i)
   +0.5*Lengthleg(i)+yokecenterb(i)],[Lengthlegyoke(i)+legsb(i)
   Lengthlegyoke(i)],'Marker','.','LineStyle','-')
line([yokesa(i)+0.5*Lengthleg(i)+yokecenterb(i) yokesa(i)
   +0.5*Lengthleg(i)+yokecenterb(i)+0.5*Lengthleg(i)],[Lengthlegyoke(i)
   u(i)],'Marker','.','LineStyle','-')
line([yokesa(i)+0.5*Lengthleg(i)+yokecenterb(i)+0.5*Lengthleg(i)
   yokesa(i)+0.5*Lengthleg(i)+yokecenterb(i)+Lengthleg(i)],
   [u(i) Lengthlegyoke(i)],'Marker','.','LineStyle','-')
line([yokesa(i)+0.5*Lengthleg(i)+yokecenterb(i)+Lengthleg(i) yokesa(i)
   +0.5*Lengthleg(i)+yokecenterb(i)+Lengthleg(i)],[Lengthlegyoke(i)
   Lengthlegyoke(i)+legsb(i)],'Marker','.','LineStyle','-')
line([yokesa(i)+0.5*Lengthleg(i)+yokecenterb(i)+Lengthleg(i) yokesa(i)
   +0.5*Lengthleg(i)+yokecenterb(i)+0.5*Lengthleg(i)],[Lengthlegyoke(i)
   +legsb(i) 2*Lengthlegyoke(i)+legsb(i)],'Marker','.','LineStyle','-')


%right bottom center yoke
pos1 = yokesa(i);
pos2 = Lengthlegyoke(i)+yokesb(i)+Lengthleg(i)+yokecenterb(i)
   +0.5*Lengthleg(i)+yokecenterb(i)+Lengthleg(i);
line(-[yokesa(i)+0.5*Lengthleg(i)-pos1-pos2 yokesa(i)-pos1-pos2],
   [Lengthlegyoke(i) u(i)],'Marker','.','LineStyle','-')
line(-[yokesa(i)-pos1-pos2 yokesa(i)+0.5*Lengthleg(i)-pos1-pos2],
   [u(i) Lengthlegyoke(i)-Lengthyoke(i)],'Marker','.','LineStyle','-')
line(-[yokesa(i)+0.5*Lengthleg(i)-pos1-pos2 yokesa(i)+0.5*Lengthleg(i)
   +yokecentera(i)-pos1-pos2],[Lengthlegyoke(i)-Lengthyoke(i)
   Lengthlegyoke(i)-Lengthyoke(i)],'Marker','.','LineStyle','-')
line(-[yokesa(i)+0.5*Lengthleg(i)+yokecentera(i)-pos1-pos2 yokesa(i)
   +0.5*Lengthleg(i)+yokecenterb(i)-pos1-pos2],[Lengthlegyoke(i)
   -Lengthyoke(i) Lengthlegyoke(i)],'Marker','.','LineStyle','-')
line(-[yokesa(i)+0.5*Lengthleg(i)+yokecenterb(i)-pos1-pos2 yokesa(i)
   +0.5*Lengthleg(i)-pos1-pos2],[Lengthlegyoke(i) Lengthlegyoke(i)],
   'Marker','.','LineStyle','-')


%right upper center yoke
line(-[yokesa(i)-pos1-pos2 yokesa(i)+0.5*Lengthleg(i)-pos1-pos2],
   [legsa(i) Lengthlegyoke(i)+legsb(i)],'Marker','.','LineStyle','-')
line(-[yokesa(i)+0.5*Lengthleg(i)-pos1-pos2 yokesa(i)+0.5*Lengthleg(i)
   +yokecenterb(i)-pos1-pos2],[Lengthlegyoke(i)+legsb(i) Lengthlegyoke(i)
   +legsb(i)],'Marker','.','LineStyle','-')
line(-[yokesa(i)+0.5*Lengthleg(i)+yokecenterb(i)-pos1-pos2 yokesa(i)
   +0.5*Lengthleg(i)+yokecentera(i)-pos1-pos2],[Lengthlegyoke(i)
```

```
    +legsb(i) legsa(i)+Lengthyoke(i)-Lengthlegyoke(i)],'Marker','.',
    'LineStyle','-')
line(-[yokesa(i)+0.5*Lengthleg(i)+yokecentera(i)-pos1-pos2 yokesa(i)
    +0.5*Lengthleg(i)-pos1-pos2],[legsa(i)+Lengthyoke(i)
    -Lengthlegyoke(i) Lengthlegyoke(i)+legsb(i)+Lengthyoke(i)],'Marker',
    '.','LineStyle','-')
line(-[yokesa(i)+0.5*Lengthleg(i)-pos1-pos2 yokesa(i)-pos1-pos2],
    [Lengthlegyoke(i)+legsb(i)+Lengthyoke(i) legsa(i)],'Marker','.',
    'LineStyle','-')


%right center leg
pos3 = Lengthleg(i)+yokecenterb(i)+Lengthleg(i)+yokecenterb(i);
line([yokesa(i)+pos3 Lengthlegyoke(i)+yokesb(i)+pos3],
    [u(i) Lengthlegyoke(i)],'Marker','.','LineStyle','-')
line([Lengthlegyoke(i)+yokesb(i)+pos3 Lengthlegyoke(i)+yokesb(i)+pos3],
    [Lengthlegyoke(i) Lengthlegyoke(i)+legsb(i)],'Marker','.',
    'LineStyle','-')
line([Lengthlegyoke(i)+yokesb(i)+pos3 yokesa(i)+pos3],[Lengthlegyoke(i)
    +legsb(i) legsa(i)],'Marker','.','LineStyle','-')
line([yokesa(i)+pos3 yokesa(i)+0.5*Lengthleg(i)+pos3],
    [legsa(i) Lengthlegyoke(i)+legsb(i)],'Marker','.','LineStyle','-')
line([yokesa(i)+0.5*Lengthleg(i)+pos3 yokesa(i)+0.5*Lengthleg(i)+pos3],
    [Lengthlegyoke(i)+legsb(i) Lengthlegyoke(i)],'Marker','.',
    'LineStyle','-')
line([yokesa(i)+0.5*Lengthleg(i)+pos3 yokesa(i)+pos3],[Lengthlegyoke(i)
    u(i)],'Marker','.','LineStyle','-')


%right end upper yoke
pos4 = Lengthleg(i)+yokecenterb(i)+Lengthleg(i)+yokecenterb(i)+yokesa(i)
    +yokesa(i);
line(-[u(i)-pos4 yokesa(i)-pos4],[u(i)+legsa(i) u(i)+legsa(i)],
    'Marker','.','LineStyle','-')
line(-[u(i)+Lengthlegyoke(i)-pos4 yokesb(i)+Lengthlegyoke(i)-pos4],
    [legsa(i)-Lengthlegyoke(i) legsa(i)-Lengthlegyoke(i)],'Marker','.',
    'LineStyle','-')
line(-[u(i)-pos4 Lengthlegyoke(i)-pos4],[legsa(i) legsb(i)
    +Lengthlegyoke(i)],'Marker','.','LineStyle','-')
line(-[yokesa(i)-pos4 yokesb(i)+Lengthlegyoke(i)-pos4],[u(i)+legsa(i)
    Lengthlegyoke(i)+legsb(i)],'Marker','.','LineStyle','-')


%right end bottom yoke
line(-[u(i)-pos4 yokesa(i)-pos4],[u(i) u(i)],'Marker','.','LineStyle',
    '-')
line(-[u(i)+Lengthlegyoke(i)-pos4 yokesb(i)+Lengthlegyoke(i)-pos4],
    [Lengthlegyoke(i) Lengthlegyoke(i)],'Marker','.','LineStyle','-')
line(-[u(i)-pos4 Lengthlegyoke(i)-pos4],[u(i) Lengthlegyoke(i)],
    'Marker','.','LineStyle','-')
line(-[yokesa(i)-pos4 yokesb(i)+Lengthlegyoke(i)-pos4],
    [u(i) Lengthlegyoke(i)],'Marker','.','LineStyle','-')


%right end leg
pos5 = Lengthleg(i)+yokecenterb(i)+Lengthleg(i)+yokecenterb(i)+yokesa(i)
    +yokesa(i);
line(-[u(i)-pos5 u(i)-pos5],[u(i) legsa(i)],'Marker','.','LineStyle',
    '-')
```

```
line(-[Lengthlegyoke(i)-pos5 Lengthlegyoke(i)-pos5],
   [u(i)+Lengthlegyoke(i) legsb(i)+Lengthlegyoke(i)],'Marker','.',
   'LineStyle','-')
line(-[u(i)-pos5 Lengthlegyoke(i)-pos5],[u(i) Lengthlegyoke(i)],
   'Marker','.','LineStyle','-')
line(-[u(i)-pos5 Lengthlegyoke(i)-pos5],[legsa(i) legsb(i)
   +Lengthlegyoke(i)],'Marker','.','LineStyle','-')

grid on
if i = =1
return
else
figure
end
end
```

Suggested Reading

1. Stevens, J., Bonn, R., Ginn, J., Gonzalez, S., Kern, G., Development and Testing of an Approach to Anti-Islanding in Utility-interconnected Photovoltaic Systems, SAND-2000-1939, Sandia National Laboratories, Albuquerque, NM, August 2000.
2. Ropp, M., Begovic, M., Rohatgi, A., Prevention of Islanding in Grid-Connected Photovoltaic Systems, *Progress in Photovoltaics Research and Applications*, 7(1), January–February 1999.
3. Bonn, R., Kern, G., Ginn, J., Gonzalez, S., *Standardized Anti-Islanding Test Plan*, Sandia National Laboratories Web page, www.sandia.gov/PV, 1998.
4. He, W., Markvart, T., Arnold, R., Islanding of Grid-Connected PV Generators: Experimental Results, Proceedings of 2nd World Conference and Exhibition on PV Solar Energy on July 6–10, Vienna, Austria, July 1998.
5. Begovic, M., Ropp, M., Rohatgi, A., Pregelj, A., Determining the Sufficiency of Standard Protective Relaying for Islanding Prevention in Grid-Connected PV Systems, *Proceedings of the Second World Conference and Exhibition on PV Energy Systems*, Hofburg Congress Center, Vienna, Austria, July 6–10, 1998.
6. Utility Aspects of Grid Interconnected PV Systems, IEA-PVPS Report, IEAPVPS T5-01: 1998, December 1998.
7. Kobayashi, H., Takigawa, K., Islanding Prevention Method for Grid Interconnection of Multiple PV Systems, *Proceedings of the Second World Conference and Exhibition on Photovoltaic Solar Energy Conversion*, Hofburg Congress Center, Vienna, Austria, July 6–10, 1998.
8. Gonzalez, S., Removing Barriers to Utility-Interconnected Photovoltaic Inverters, *Proceedings of the 28th IEEE PV Specialists Conference*, Anchorage, AK, September 2000.
9. Van Reusel, H., et al., Adaptation of the Belgian Regulation to the Specific Island Behaviour of PV Grid-Connected Inverters, *Proceedings of the 14th European Photovoltaic Solar Energy Conference and Exhibition*, Barcelona, Spain, pp. 2204–2206, 1997.
10. Häberlin, H., Graf, J., Islanding of Grid-Connected PV Inverters: Test Circuits and Some Test Results, *Proceedings of the Second World Conference and Exhibition on Photovoltaic Solar Energy Conversion*, Vienna, Austria, pp. 2020–2023, 1998.
11. Ropp, M., Begovic, M., Rohatgi, A., Analysis and Performance Assessment of the Active Frequency Drift Method of Islanding Prevention, *IEEE Transactions on Energy Conversion*, 14(3), 810–816, 1999.
12. Ambo, T., Islanding Prevention by Slip Mode Frequency Shift, *Proceedings of the IEA PVPS Workshop on Grid-interconnection of Photovoltaic Systems*, September 1997.
13. Kitamura, A., Islanding Prevention Measures for PV Systems, *Proceedings of the IEA PVPS Workshop on Grid-Interconnection of Photovoltaic Systems*, September 1997.
14. Nanahara, T., Islanding Detection—Japanese Practice, *Proceedings of the IEA PVPS Workshop on Grid-Interconnection of Photovoltaic Systems*, September 1997.
15. Häberlin, H., Graf, J., Beutler, C. Islanding of Grid-Connected PV Inverters: Test Circuits and Test Results, *Proceedings of the IEA PVPS Workshop on Grid-Interconnection of Photovoltaic Systems*, September 1997.

16. Bonn, R., Ginn, J., Gonzalez, S., Standardized Anti-Islanding Test Plan, Sandia National Laboratories, January 26, 1999.
17. Yuyuma, S., et al., A High-Speed Frequency Shift Method as a Protection for Islanding Phenomena of Utility Interactive PV Systems, *Solar Energy Materials and Solar Cells (1994)*, 35, 477–486.
18. Kitamura, A., Okamoto, M., Yamamoto, F., Nakaji, K., Matsuda, J., Hotta, K., Islanding Phenomenon Elimination Study at Rokko Test Center, *Proceedings of the First IEEE World Conference on Photovoltaic Energy Conversion (1994)*, Part 1, pp. 759–762.
19. Kobayashi, H., Takigawa, K., Islanding Prevention Method for Grid Interconnection of Multiple PV Systems, *Proceedings of the Second World Conference and Exhibition on Photovoltaic Solar Energy Conversion*, Hofburg Congress Center, Vienna, Austria, July 6–10, 1998.

Appendix B

Standards, Codes, User's Guides, and Other Guidelines

S1. ANSI/NFPA 70, *The National Electrical Code*, 2002, National Fire Protection Association, Batterymarch Park, Quincy, MA, September 2001.

S2. UL1741, *UL Standard for Safety for Static Converters and Charge Controllers for Use in Photovoltaic Power Systems*, Underwriters Laboratories, First Edition, May 7, 1999, Revised January 2001.

S3. IEEE Standard 929-2000, *IEEE Recommended Practice for Utility Interface of Photovoltaic (PV) Systems*, Sponsored by IEEE Standards Coordinating Committee 21 on Photovoltaics, IEEE, New York, April 2000.

S4. DIN VDE 0126:1999, Automatic Disconnection Facility for Photovoltaic Installations with a Nominal Output <4.6 kVA and a Single-Phase Parallel Feed by Means of an Inverter into the Public Grid (*German National Standard for Utility Interconnection of Photovoltaic Systems*).

S5. *Small Grid-Connected Photovoltaic Systems*, KEMA Standard K150, 2002.

S6. *Guidelines for the Electrical Installation of Grid-Connected Photovoltaic (PV) Systems*, Dutch guidelines to comply with NEN1010 (safety provisions for low voltage installations), EnergieNed, and NOVEM, December 1998.

S7. *Supplementary Conditions for Decentralized Generators—Low Voltage Level*, Dutch Guidelines to comply with NEN1010 (safety provisions for low voltage installations), EnergieNed and NOVEM, April 1997.

S8. JIS C 8962:1997, *Testing Procedure of Power Conditioners for Small Photovoltaic Power Generating Systems*, Japanese Industrial Standard, 1997. EN61277, *Terrestrial Photovoltaic (PV) Power Generating Systems— General and Guide*.

S9. ÖNORM/ÖVE 2750, *Austrian Guideline for Safety Requirements of Photovoltaic Power Generation Systems*.

S10. AS3000, *Australian Guidelines for the Grid Connection of Energy Systems via Inverters*.

S11. AS4777, *Grid Connection of Energy Systems via Inverters,* Proposed joint Australia/New Zealand Consensus Standard, May 2005.

S12. DIN VDE 0100 Teil 712 *Photovoltaische Systeme, Amendment to Germany's Basic Electrical Safety Code.*

S13. G77, *UK Standard for Interconnection of PV and Other Distributed Energy,* Generation, Standard under development, expected completion, January 2002.

S14. CSA F381, *Canadian Standard for Power Conditioning Systems.*

S15. ASTM Standard E 1328, *Standard Terminology Relating to Photovoltaic Solar Energy Conversion.*

S16. EN61277, *Terrestrial Photovoltaic (PV) Power Generating Systems—General and Guide.*

S17. IEEE/ANSI Standard C37.1-1994, *IEEE Standard Definition, Specification, and Analysis of Systems Used for Supervisory Control, Data Acquisition, and Automatic Control.*

S18. IEEE Standard 519-1992, *IEEE Recommended Practices and Requirements for Harmonic Control in Electric Power Systems (ANSI),* IEEE, New York.

S19. ESTI No. 233.0690: Photovoltaische Energieerzeugungsanlagen— Provisorische Sicherheitsvorschriften (Photovoltaic Power Generating Systems—Safety Requirements), Adopted by Switzerland, 1990 draft.

S20. VSE Sonderdruck Abschnitt 12, *Werkvorschriften über die Erstellung von elektr. Installation* Elektrische Energieerzeugungsanlagen, Completes the VSE2.8d-95, Adopted by Switzerland, 1997.

S21. International Standard IEC 62116 DRAFT, *Testing Procedure of Islanding Prevention Measures for Grid Connected Photovoltaic Power Generation Systems,* International Electrotechnical Commission.

S22. IEC 61836: 1997, *Solar Photovoltaic Energy Systems—Terms and Symbols.*

S23. Considerations for Power Transformers Applied in Distributed Photovoltaic (DPV)—Grid Application, DPV-Grid Transformer Task Force Members, Power Transformers Subcommittee, IEEE-TC, Hemchandra M. Shertukde, Chair, Mathieu Sauzay, Vice Chair, Aleksandr Levin, Secretary, Enrique Betancourt, C. J. Kalra, Sanjib K. Som, Jane Verner, Subhash Tuli, Kiran Vedante, Steve Schroeder, Bill Chu, white paper in preparation for final presentation at the IEEE-TC conference in San Diego, CA, April 10–14, 2011.

S24. C57.91: IEEE Guide for Loading Mineral-Oil-Immersed Transformers, 1995, Correction 1-2002.

S25. C57.18.10a: IEEE Standard Practices Requirements for Semiconductor Power Rectifier Transformers, 1998 (amended in 2008).

S26. C57.110: IEEE Recommended Practice for Establishing Liquid-Filled and Dry-Type Power and Distribution Transformer Capability When Supplying Non-Sinusoidal Load Current, 2008.

S27. C57.116: IEEE Guide for Transformers Directly Connected to Generators, 1989.

S28. C57.129: IEEE Standard for General Requirements and Test Code for Oil-Immersed HVDC Convertor Transformer, 1999 (2007–Approved).

S29. Standard 1547.4: Draft Guide for Design, Operation and Integration of Distributed Resource Island Systems with Electric Power System (only 1547.1 is there), 2005.

S30. UL 1741: A Safety Standard for Distributed Generation, 2004.

S31. Buckmaster, David, Hopkinson, Phil, Shertukde, Hemchandra, Transformers Used with Alternative Energy Sources—Wind and Solar, Technical presentation, April 11, 2011.

S32. Standard 519: Recommended Practices and Requirements for Harmonic Control in Electrical Power Systems, 1992.

Index

FIGURE 2.1
Typical power transformer used in a solar inverter type application. (Courtesy of Power Distribution, Inc. dba Onyx Power.)

(a)

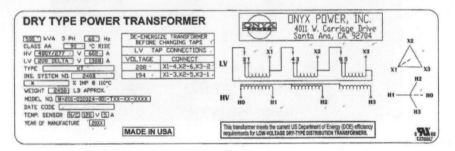

(b)

FIGURE 2.2
(a) DPV-GT dry-type (fiberglass insulated) for solar application. (Courtesy Power Distribution, Inc. dba Onyx Power.) Note the simple winding configuration on the secondary of the delta/wye neutral grounded configuration in a frontal view. (b) Nameplate details of DPV-GT dry-type (fiberglass insulated) for solar application for transformer in Figure 2.2a. (Courtesy of Power Distribution, Inc. dba Onyx Power.)

FIGURE 2.3
Many transformers are pad-mount type.

FIGURE 2.4
Dry-type CSP power transformers in solar PV grid application. (Courtesy of Diagnostic Devices Inc.)

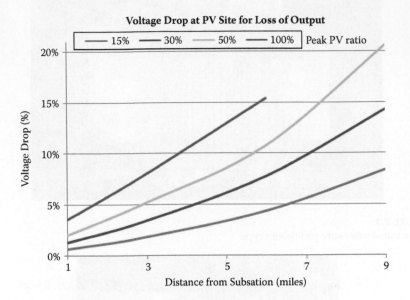

FIGURE 3.1
Percentage of voltage drops as a function of distance from substation and DPV-GT.

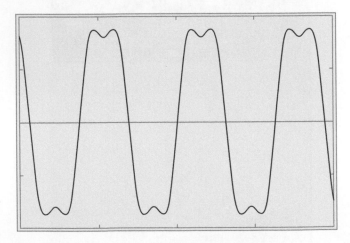

FIGURE 4.1
Fundamental frequency with an in-phase third-harmonic frequency.

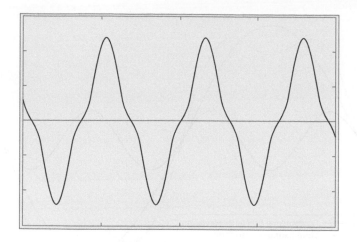

FIGURE 4.2
Fundamental frequency with an out-of-phase third-harmonic frequency.

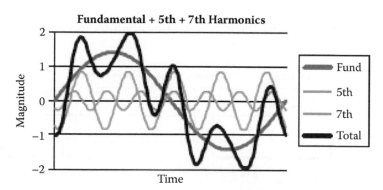

FIGURE 4.3
Fundamental and odd harmonics (fifth and seventh).

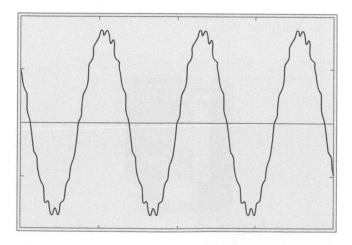

FIGURE 4.5
Distorted sinusoid containing all the harmonics.

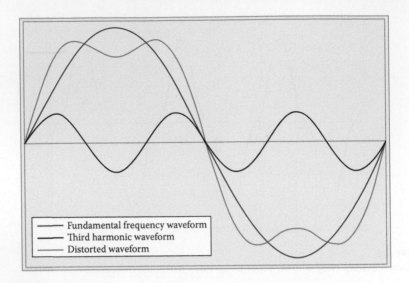

FIGURE 4.6
Synthesized waveforms with fundamental and third in-phase harmonic component.

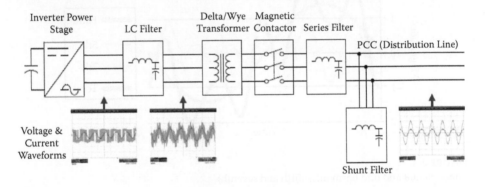

FIGURE 4.8
Circuit for harmonic mitigation.

FIGURE 7.1
Protection relay for a fault on circuit adjacent to a DPV circuit.

FIGURE 7.2
Voltage relay for a fault on circuit adjacent to a DPV circuit.

FIGURE 7.3
Feeder protection relay for a fault on circuit adjacent to a DPV circuit.

FIGURE 7.4
Biased differential relay for a fault on circuit adjacent to a DPV circuit.

FIGURE 12.1
Sun-Solar panel-inverter DPV-GT feeder lines.

FIGURE 12.6
Inverter for grid-connected PV.

FIGURE 12.7
Example of large, three-phase inverter for commercial and utility-scale grid-tied PV systems.

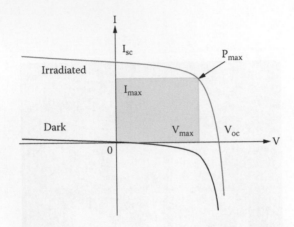

FIGURE 12.8
Current-voltage characteristics of a solar cell: the area of the rectangle gives the output power; and P_{max} denotes the maximum power point.

FIGURE 12.9
A solar micro-inverter.

FIGURE 12.10
Released in 1993, Mastervolt's Sunmaster 130S was the first true micro-inverter.

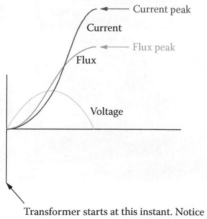

FIGURE 12.11
Another early micro-inverter, the OK4E-100 (1995) (E for European, 100 for 100 W).

Current peak

Current

Flux peak

Flux

Voltage

Transformer starts at this instant. Notice
that the voltage is zero at this instant.

FIGURE 13.2
Peak current curves for transformer protection.

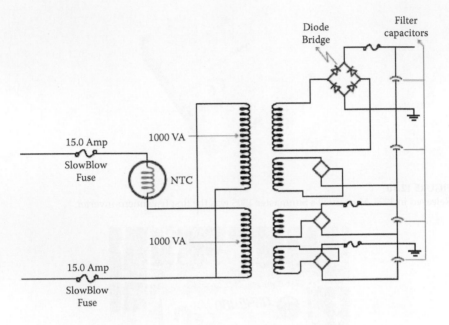

FIGURE 13.3
Typical circuit using a thermistor for protection.

FIGURE 15.2
Foil winding machine with foil being wound.

FIGURE 15.3
Spools, which carry the aluminum foil, assembled at the back of the foil winding machine.

FIGURE 15.4
Rectangular aluminum foil, layer or helitran winding.

FIGURE 15.18
Coil clamping machine and fixture.

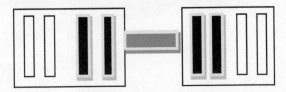

FIGURE 15.22
Magnetic circuit design with air-gap in the central limb.

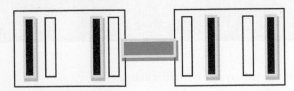

FIGURE 15.23
Windings arranged in interleaved fashion to reduce losses by fringing.

FIGURE 16.1
Four-channel PD diagnostic device. (Courtesy of Diagnostic Devices Inc.)

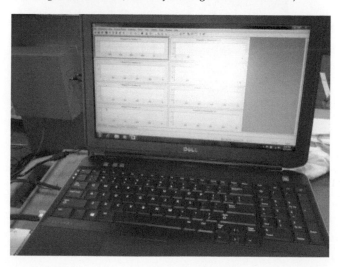

FIGURE 16.2
Data screen of a four-channel PD diagnostic device. (Courtesy of Diagnostic Devices Inc.)

FIGURE 16.3
1600 kVA dry-type DPV-GT being tested with a four-channel briefcase PD diagnostic device.
(Courtesy of Diagnostic Devices Inc.)

Printed and bound by CPI Group (UK) Ltd, Croydon, CR0 4YY

18/10/2024

01776262-0006